The Conquest of Tuberculosis

Selman A. Waksman

The Conquest of
TUBERCULOSIS

UNIVERSITY OF CALIFORNIA PRESS
BERKELEY AND LOS ANGELES 1964

University of California Press
Berkeley and Los Angeles, California

Cambridge University Press
London, England

© 1964 by The Regents of the University of California
Library of Congress Catalog Card Number: 64-21065

"If I had tuberculosis . . .

> *this idea, formerly terrifying, no longer makes any-
> one tremble . . . antibiotics have appeared, sanatoria
> have disappeared; as far as the public is concerned the
> problem is solved, the disease has been conquered."*

INTERNATIONAL UNION AGAINST TUBERCULOSIS, JANUARY, 1962

TO THE SURGEONS AND PHYSICIANS OF THIS AND OTHER
COUNTRIES, WHO HAVE TAKEN ADVANTAGE OF THE DIS-
COVERY OF ANTIBIOTICS, LIKE THE ONE DESCRIBED IN THIS
BOOK, TO IMPROVE THE HEALTH OF MAN AND TO CURE
INFECTIOUS DISEASES, THIS BOOK IS DEDICATED.

PREFACE

This book is not a textbook. Neither is it a treatise dealing with the nature of tuberculosis and methods of treatment. It is the story of a disease—an ancient disease that has afflicted the human race since man's beginnings, and how it is finally being brought under control. Tuberculosis has not yet been eradicated, and the question is often raised whether it can ever be, but it can be controlled. No one who becomes afflicted by any one of the different forms of this disease need die from it at the present time.

I am not a medical authority, I make no claim to any profound knowledge of the disease itself, of its various manifestations, or of specific methods of treatment. I feel justified, however, on the basis of observations during the past twenty years of the gradual disappearance of the disease and of the evacuation of the sanatoria and hospitals dedicated to it, in attempting to tell the story of how this all came about.

The literature on the subject is so vast that only limited sources could be consulted. I have drawn freely upon some, especially the more recent publications, and wish to express here my indebtedness. Credit is, of course, given to each citation in its proper place.

I am grateful to all those who have read parts of the manuscript and who have freely made various constructive suggestions. I want to express special appreciation to Dr. Floyd Feldmann, Dr. Esmond Long, and Dr. H. Corwin Hinshaw, and to Professor G. Canetti and Dr. Noël Rist of the Pasteur Institute in Paris, for their many critical comments and corrections. Full advantage has been taken of these. I am indebted to all the authors whose work has been cited freely in the text. I also want to express my sincere appreciation to Mrs. R. Nehlig, who has typed and retyped the manuscript and helped in the verification of the many quotations. I should also like to extend my thanks to Mr. Jesse Phillips for editing the manuscript and to members of the staff of the University of California Press for help during the production of the work.

I also want to express particular appreciation to the surgeons, physicians, and nurses of the Sanatorio Americano, the Medical School, and the School of Nursing, Montevideo, Uruguay, who saved my own life when I fell ill recently on a lecture visit to that far-off city, and made possible my speedy recovery from a dreadful affliction, enabling me to regain my normal health and complete my life work.

This volume is being published on the twentieth anniversary of the introduction of streptomycin into clinical uses. The motto engraved on the monument of Dr. Trudeau for the benefit of all tuberculosis sufferers who came to the Saranac Lake Sanatorium —"To cure sometimes, to relieve often, to comfort always"—may now well be changed to read: "To save all from the formerly dreadful disease—tuberculosis."

SELMAN A. WAKSMAN
Professor Emeritus
Rutgers, The State University

February 15, 1964

ACKNOWLEDGMENTS

Acknowledgment of permission to reprint is gratefully extended to: Doubleday and Company, Inc., for passages from *An Autobiography* by E. L. Trudeau; Paul B. Hoeber, Inc., for passages from *Tuberculosis* by G. B. Webb; Lea and Febiger, Inc., for passages from *Tuberculosis* by E. R. Baldwin and *Pulmonary Tuberculosis* by M. Fishberg; Little, Brown and Company, for passages from *The White Plague: Tuberculosis, Man and Society* by René and Jean Dubos; The Williams and Wilkins Co. for passages from *Fighters of Fate* by J. A. Myers and *The Story of Clinical Pulmonary Tuberculosis* by L. Brown.

CONTENTS

INTRODUCTION

Tuberculosis has been known since ancient times, under a variety of names, in different historical periods, and in the various countries of the world. It has often been confused with certain other diseases, but its basic nature, namely, its wasting effect and its destructive character, has always been recognized. Until very recently it was known as the number-one killer of the human race. In the words of Ashburn (1947), tuberculosis "transcends all maladies in the total number of its victims and the cost to society." Its destructive effects have varied greatly throughout human history.

Prehistoric man suffered from tuberculosis, as shown by various neolithic skeletal remains. The disease was familiar to the most ancient civilizations, as can be seen from the inscriptions on Babylonian tablets and on other relics of antiquity. It was clearly recognized by the ancient Greeks. Definite evidence of the affliction that tuberculosis has brought down upon man can be gained from various ancient monuments.

In the Bible, Jehovah threatened the people of Israel: ". . . I will even appoint over you terror, consumption, and the burning ague, that shall consume the eyes, and cause sorrow of heart . . ." (Lev. 26:16.)

The disease did not attack any one particular organ in the human body, but inflicted its damage upon a variety of organs and tissues. This in itself is sufficient to explain why so many different names have been applied to it. "Phthisis" and "scrofula" were among the most common; the "great white plague" was another; "consumption" and "wasting disease" were used until recent years.

The actual name "tuberculosis" was introduced during the first half of the nineteenth century to designate a group of diseases characterized by the production of tubercles in different parts of the body. More recently, it came to mean any infection caused by tubercle bacilli, whether or not tubercles could be found in the infected organs. Men and women at all stages of life, but primarily between the ages of fifteen and thirty-five, were subject to the disease. It afflicted not only man himself, but also his cattle and his other domestic animals. The disease thus played an important role not only in man's health, but also in his economy.

The infectious nature of tuberculosis was not generally recognized until a century or so ago, although it was suspected even in olden times, for afflicted individuals were isolated in some of the more advanced ancient communities. It was believed that over centuries of exposure to the disease those who could not develop a protective immunity were killed off and that this rigorous selective process gradually raised the general resistance of mankind to the disease. Whether this is true or not, we do know that when the disease was rampant and practically everybody was infected at some time or other, the death rate was far from 100 per cent.

There was no certain cure for this destructive disease, especially of certain of its forms or manifestations, such as generalized (miliary) tuberculosis or tuberculous meningitis (on the covering of the brain and spinal cord). The rapidly progressive form of the disease was called "galloping consumption." Thomas Young wrote, nearly a century and a half ago (1815), that not one patient in a thousand recovered from this disease without medical help; even then the best possible care could hardly save the life of one person in a hundred.

During the past two decades my own life became intimately interwoven with the control of infectious disease in general and tuberculosis in particular. I have never suffered from any form of this disease, nor have any members of my immediate family. Not being a medical authority, I have never examined patients suffering from it. I have never even isolated the organism (generally known under the name of *Mycobacterium tuberculosis*) from a diseased patient. I have, nevertheless, been "catapulted" into contact with it through my work on the microbes inhabiting the soil and through my study of the production by some of these microbes of chemical compounds, known as antibiotics. These substances proved to have a destructive effect upon various disease-producing bacteria, including the tuberculosis organism. They have thus served to deliver the first blow against this ancient foe of man.

My own contact with tuberculosis came about in the following manner. In 1932 I was asked by the Chairman of the Medical Division of the National Research Council, Dr. Stanhope Bayne-Jones, and by the Director of Research of the National Tuberculosis Association, Dr. William C. White, to undertake a study of the fate of the tuberculosis organism in soils and in water basins, since I was deeply engaged at that time in a survey of the microbial populations of these two natural habitats. After an exhaustive study, carried on in my laboratory largely in collaboration with one of my graduate students, we succeeded in demonstrating that this bacterium gradually disappears in such habitats and that certain other microbes saprophytic in nature appear to be at least partly responsible for its disappearance. On the completion of this study in 1935, I was not yet prepared to take advantage of these findings and did not attempt to determine the exact mechanisms involved in the process of destruction of the *Mycobacterium* in the natural environment. Neither was I prepared to undertake such a study a year or so later when a test-tube culture of the avian form of the tuberculosis organism, apparently killed by a contaminating mold, was brought to me by a colleague from the Poultry Department of my university.

Another seven or eight years were to pass before the broadening knowledge of antibiotics suggested the desirability of testing the effect of these substances upon the organism causing tuberculosis in man and in animals. In 1943, two of my assistants and I isolated a new antibiotic from the culture of a soil microbe. The latter belonged to a group of bacteria known as actinomycetes. Its scientific name was *Streptomyces griseus;* the new antibiotic was, therefore, designated *streptomycin.* We soon demonstrated that this antibiotic has the capacity to inhibit the growth of *Mycobacterium tuberculosis* in test tubes and in flask cultures. This was immediately followed by the work of Drs. H. Corwin Hinshaw and William Feldman of the Mayo Clinic, who demonstrated the remarkable effect of this antibiotic in inhibiting the development of tuberculosis in experimental animals, and later in man. Thus began a revolution in the treatment of this disease. The long-hoped-for "cure" for this ancient foe of man was now within reach. Chemotherapy of tuberculosis was now a long step nearer realization. The outlook for complete control of the disease was promising.

This book is an attempt to survey the historical background of tuberculosis and the progress that has been made toward its control and final eradication.

« Part I »

EARLY HISTORY OF TUBERCULOSIS.
EMPIRICAL METHODS OF TREATMENT

«1»

Since Time Immemorial Man Has Been Afflicted by a Dreadful Disease, Variously Known as Phthisis, Scrofula, Consumption, and Tuberculosis

> *The Lord shall smite thee with a consumption, and with a fever, and with an inflammation, and with an extreme burning, and with the sword, and with blasting, and with mildew; and they shall pursue thee until thou perish.* DEUT. 28:22

Tuberculosis, or consumption, as the disease was termed in earlier days, has been known since ancient times as one of the greatest scourges of man, afflicting particularly young people and leading to their early death; it also attacked both domestic and wild animals. The disease has been known also under other names, such as phthisis, scrofula, tabes, bronchitis, inflammation of the lungs, hectic fever, gastric fever, and lupus. Frequently tuberculosis was confused with other afflictions, so that even the Biblical "consumption" may represent not only true tuberculosis, but also other ailments.[1] Tuberculosis in a proper sense refers to a disease condition caused by infectious agents known as tuberculosis bacteria or tubercle bacilli.

[1] The word "consumption" is employed in the Bible to translate *schachepheth*, which is also the modern Hebrew word for tuberculosis.

There are numerous references in medical and nonmedical literature to tuberculosis under a variety of other designations, many of which are misleading. The word "phthisis" was used by the ancient Greeks because of the wasting effect upon the body, owing to the profuse diarrhea which takes place in the terminal stage of the disease and the skin-drenching sweats occurring in the middle and later stages. This word was gradually replaced by the word "consumption," introduced by English-speaking people, which expresses the burning up and destruction of the body. The word "tuberculosis" (derived from the Latin *tuberculum,* or lump) came into general use only in very recent years. When the tuberculosis organism finds lodgment in the tissues, a small lump is formed in the first stage of the tuberculous process. When a number of small lumps run together, a larger lump results and may assume the size of a nut or an egg.

Castiglioni, in his *History of Tuberculosis* (1933), states: "The opinions regarding tuberculosis held by ancient peoples are confusing and uncertain. The picture is brilliantly illuminated by the appearance of the gigantic figure of Hippocrates (460-377 B.C.), who considered [tuberculosis] the most grave of all the diseases, the most difficult to cure, and the most fatal. He regarded phthisis or pulmonary tabes as a disease following ulceration or suppuration of the lungs [and] this concept is particularly evident in the search of etiological causes in pulmonary tuberculosis."

According to Lawrason Brown (1941), "the Indo-Aryans (1500 B.C.) thought that the disease developed from over-fatigue, sorrow, fasting, pregnancy and chest wounds. It was impure and no Brahmin was allowed to marry into a family where it existed . . . The treatment was really hygienic-dietetic, looking toward improving the general condition of the patient [and] walking, carriage rides, life in the mountains or in goat stables, were ordered . . . Among the Greeks, patients with phthisis were cared for in the temples which were often found in salubrious climates. Here they were dieted, given gymnastics and milk, and even sent on sea voyages. Any disease with emaciation was diagnosed by Hippocrates as phthisis, for phthisis meant really loss of weight . . . The physician was advised to avoid the advanced patients as they were sure to die and so injure his reputation . . . The idea of contagion was general among the laity and upheld by the greatest of their scientists, Aristotle. Belief in heredity was common . . . Aretaeus, the Cappadocian, a Greek from Asia Minor, stated that while the lung was not sensitive the pleura was highly so, and is said to have differentiated empyema from phthisis."

Moorman (1940) states that ". . . the serious study of history justifies the belief that tuberculosis may have been the first-born of the mother of pestilence and disease. Exhumed skeletons of prehistoric periods bear the marks of tuberculosis. Thus we see that before the time of recorded history tuberculosis left an ineradicable record of its ravages. The Code of Hammurabi, written before 2,000 B.C., indicates a knowledge of the disease."

Hippocrates, in the fifth century B.C., stated that "Phthisis was really the most prevalent of the diseases which prevailed at that time and the only one which killed the patients" (Flick, 1925). Tuberculosis has thus been recognized as the great scourge of all civilized peoples as far back as history records anything about human affairs. It resulted in much suffering. Although the number of its victims has been greatly reduced in modern times as a result of the discovery of the cause of the disease and greater appreciation of the factors that bring it about, it has remained a major cause of suffering well into this century.

An early Roman physician, Caelius Aurelianus (quoted in Drabkin, 1950), gave a masterly description of pulmonary tuberculosis:

"The symptoms of the disease are as follows: there is a latent fever, which generally begins toward the end of the day and is relieved by the coming of the new day; this is accompanied by much coughing at the beginning and the end of the night, with the discharge of sanious sputa . . . The voice is either hoarse or high pitched, breathing is difficult, cheeks flushed, and the rest of the body ashen colored. The eyes have a worn appearance, and the patient is emaciated, a fact more obvious from the appearance of his naked body than from his countenance. In some cases there is a hissing sound or wheezing in the chest; and, as the disease spreads, there is sweating in the upper parts down to the end of the chest. There is a loss of appetite for food, or else a considerable appetite; the patient also suffers thirst . . . The tops of the fingers become thick, and there is a hooking of the nails . . . Many subject the purulent sputa to diagnostic tests. Thus they place the phlegm over hot coals and note its odor when it has burned; for a foul odor always characterizes the product of physical decomposition."

Although the writers of the Old Testament did not fully appreciate the nature of consumption, we find numerous references to the isolation and quarantine of those ill with leprosy: "And the leper in whom the plague is, his clothes shall be rent, and his head bare, and he shall put a covering upon his upper lip, and shall cry, Unclean, unclean. All the days wherein the plague shall be in him he shall be defiled; he is unclean; he shall dwell alone; without the camp shall his habitation be . . . He shall therefore burn that garment, whether warp or woof, in woolen or in linen, or any thing of skin, wherein the plague is: for it is a fretting leprosy; it shall be burnt in the fire." (Lev. 13:45-46, 52.)

Dubos and Dubos state in their book *The White Plague* that "Tuberculosis . . . appeared to be so constantly and universally present that there was a tendency to regard it as an act of God, affecting both the rich and the poor and against which little action was possible. Moreover, the disease was relatively slow in its course and did not produce repulsive lesions or symptoms. Pallor, emaciation, weakness, and even cough were actually considered fascinating in early Victorian days."

With the passing of time, the concept of tuberculosis has changed. The doctor at first recognized it by its ravages, by the symptoms produced and the damage caused to the human organism; to the sufferer, however, it was a struggle with a relentless enemy.

According to Flick, "In individuals in whom the tubercle bacillus grows meagerly, in whom it has produced but slight toxemia and in whom it has set up no serious changes in the tissues, it not only may give no discomfort but may stimulate the functional activity of those organs of the body which have to do with the enjoyment of life. . . . In such cases the local injury is not enough to interfere with the functions of the organ invaded and the slight toxemia stimulates instead of depresses, thereby enabling the individual to do his chosen work better than he could do it otherwise and giving him ambition to accomplish things which otherwise he might not have. This is the stimulation which sometimes has enabled poets to sing more sweetly, musicians to give finer music to the world, artists to portray life on their canvasses more clearly, and patriots and statesmen to devote themselves more heroically to the good of their country and of the people. It is what has put that sweetness into human life which places the classical halo around the head of the individual who, in the language of the past, was described as having the phthisical diathesis."

Description of Disease by
Some of the Early Great Clinicians

Thomas Willis wrote in 1672 (as quoted by Flick): "Among diseases of the chest, phthisis or consumption . . . by right holds the first place; for there exists nothing more frequent or more difficult to cure. But really all the other diseases of the chest, bad or in a measure curable, lead to this disease as small streams lead into a big lake so that they finally terminate in phthisis and naturally in the end take this name. In fact, this word consumption in its proper significance not only includes phthisis but atrophy or wasting of the solid parts with loss of strength; and this affection in all its frequency leads to the same final softening or ulceration of the lungs; indeed, wherever it may arise, whether in the evils already mentioned or merely from the blood or in some fault of the nervous juice or in a thinning of the whole body or in wasting."

Sylvius, in his *Opera Medica* of 1679 (as cited by Flick) first described tubercles as constant and characterized lesions in the lungs and other parts of the body of tuberculosis sufferers. He also noted their progression into cavities and their suppuration to give rise to "ulcers." The tubercles were believed to be swollen, diseased lymph glands. A similarity of the tubercles to the nodules observed in the necks and other parts of those suffering from scrofula was also noted.

Richard Morton wrote in 1689, according to Flick, the most pretentious book upon the subject of consumption, entitled *Phthisiologia.* He said that "A consumption of the lungs is an universal wasting of the parts of the body caused by distemper of the lungs, such as a stuffing, swellings, inflammation and exulceration of them and thereupon it is attended with a cough, difficulty of breathing and other symptoms of the breast, and accompanied with a fever, which at first is slow, and hectic, afterwards inflammatory and at last putrid and intermitting. . . . When the inflamed swellings of the lungs begin to be apostemes, . . . this inflammatory fever is changed into a putrid intermitting fever." He recognized early stages in phthisis, also its recurrences, and suggested that most men harbored the disease at one time or another.

Sydenham (1624-1689), frequently spoken of as the English Hippocrates, anticipated our present concern with air pollution: "The frequency of consumptions in London is for that we live here in a perpetual mist, the sun not being powerful enough to dissipate the clouds, and with this mist are mixed the fumes that arise from the several trades managed here but especially the sulphur and fumes of sea-coals with which the air is repleated and these, being sucked into our lungs and insinuating into the blood itself, give occasion for a cough . . . The second sort of consumption is laid in quite contrary season, viz., in the beginning of the summer; for about that time a spitting of blood happens often to such young men whose blood is weak but hot and sharp after violent exercise or a debauch of drinking . . .

A third sort of phthisis happens in the end of a fever, when the febrile matter is discharged upon the lungs and so, in the place of the essential fever, there succeeds a hectic . . . and, not very long after, a *diarrhoea lethalis,* for they soon die of this sort of phthisis." (Quoted in Cummins, 1949.)

Manget observed in 1700, while performing an autopsy on a youth, that the tubercles were so small as to resemble "millet seed" and were present over the entire surface of the lungs, "front and back and in between the large lobes, in the interstitial tissue," the lungs being "strewn with white bodies rather hard, of the size of a millet seed, of the white poppy, and some of the size of a hemp seed, closely joined together, scarcely leaving any part of the lung free from them" (after Flick). This description of the disseminated disease led to the designation of "miliary" tuberculosis.

Marten, in his classic book *A New Theory of Consumption,* in 1720, defined tuberculosis as follows: "Custom has now so much prevailed with physicians that whenever they speak of a consumption it is generally and more especially taken for a phthisis or that consumption of the body which has its rise from an ulceration of the lungs. A phthisis or consumption of the lungs may be very justly defined to be a wearing away or consuming of all the muscular or fleshy parts of the body, accompanied with a cough, purulent spitting, hectic fever, shortness of breath, night sweats, etc. . . . The divine Hippocrates and from him several others tell us that persons with a fine contexture, tender, and who have a small shrill voice, thin clear skin, a long neck, narrow breast, depressed or strait chest and whose shoulder blades stick out are of all others most subject to consumption; and this is in some measure confirmed by experience, *but must not be taken as a general rule because we often find robust and strong men fall into this distemper and such weakly tender persons as above described many times exempted from it* . . . Consumptive people are likewise generally observed to be very full of spirit, hasty and of a sharp and ready wit . . . People between the age of eighteen and thirty-five are much more subject to a consumption than those who are either younger or older *yet this must not be taken as a rule neither.*" (Quoted in Cummins, 1949.)

NATURE AND DESCRIPTION OF TUBERCULOSIS

As we come down to modern times, we find a much broader concept of the nature of the various forms of tuberculosis, their causation, the effect of environment, and methods of treatment.

Huber described in 1906 the clinical course of those who suffered from this disease: "Those thus afflicted become progressively very weak and very much emaciated. Their hearts beat rapidly and they are apt to have a pink flush on their cheeks, which is quite unlike the blush of health, but which is in reality an indication of the fever that is consuming them. The rest of their faces is very placid and thin and is suffused with a clammy sweat. Their cheek-bones are prominent; and their eyes have a quite unnatural brilliancy, seeming large and beautiful. But their lustre is not of health,—rather of disease, and too often of death. And the consumptive spits blood sometimes, and is short of breath, and has a persistent, hacking cough, that harasses him dreadfully, and will not let him rest."

The behavior of the consumptive has been described in a similar vein by Fishberg (1932): "His bright eyes with dilated pupils, which are at times contracted unilaterally, the flushing cheeks, the keen intellect which is so often met with among those who before the onset of the disease were rather dull in this respect, coupled with a flickering intelligence which brightens up suddenly for a few hours, but is soon followed by mental depression or fatigue, bear close resemblance to the average person who is under the influence of moderate doses of alcohol, or a narcotic drug. In tuberculous patients, particularly young talented individuals, it is noted that for weeks or months, now and then, they display enormous intellectual capacity of the creative kind. Especially is this to be noted in those who are of the artistic temperament, or who have a talent for imaginative writing. They are in a constant state of nervous irritability, but despite the fact that it hurts their physical condition, they keep on working and produce their best work . . ."

In his discussion of the diseases of the lungs Babcock (1907) states: "The one peculiarity of the consumptive which probably strikes the observer most forcibly is his hopefulness, the *spes phthisicorum* of the ancients. This is not usually seen, or at least is not pronounced, in the beginning of the disease, but late in its course, when it is only too apparent to his friends that death is not far off, the consumptive is possessed with a belief in his speedy recovery. He not only talks hopefully of his condition, but actually makes plans for the future which to his friends are absurd and distressing. It is this sanguine expectancy which makes the bedridden consumptive so ready to undertake journeys to some vaunted resort."

One of the great American experts on tuberculosis, who himself became afflicted by the disease, Trudeau (fig. 1), describes in his autobiography (1916) what he felt when told about his affliction: "I think I know something of the feelings of the man at the bar who is told that he is to be hanged on a given date, for in those days, pulmonary consumption was considered as absolutely fatal. I pulled myself together, put as good a face on the matter as I could, and escaped from the office after thanking the doctor for his examination. When I got outside . . . I felt stunned. It seemed to me that the world had grown suddenly dark. The sun was shining it is true, and the street was filled with the rush and noise of traffic, but to me the world had lost every vestige of brightness. I had consumption—that most fatal of diseases! Had I not seen it in all its horrors in my brother's case? It meant death and I had never thought of death before. Was I ready to die? How could I tell my wife whom I had just left in unconscious happiness with the little baby in our new home? And my rose-coloured dreams of achievement and professional success in New York! They were all shattered now and, in their place, only exile and the inevitable end remained."

It is to the novelists and poets that we must look, however, for detailed descriptions of tuberculosis, although they often called it by some other name.

FIG. 1. E. L. Trudeau.

Charles Dickens, in his novel *Nicholas Nickleby,* wrote in poetic prose quite foreign to professional medical works: "There is a dread disease which so prepares its victim, as it were, for death; which so refines it of its grosser aspect, and throws around familiar looks, unearthly indications of the coming change—a dread disease, in which the struggle between soul and body is so gradual, quiet, and solemn, and the result so sure, that day by day, and grain by grain, the mortal part wastes and withers away, so that the spirit grows light and sanguine with its lightening load, and, feeling immortality at hand, deems it but a new term of mortal life; a disease in which death takes the glow and hue of life, and life the gaunt and grisly form of death; a disease which medicine never cured, wealth warded off, or poverty could boast exemption from; which sometimes moves in giant strides, and sometimes at a tardy pace, but, slow or quick, is ever sure and certain."

Dumas the younger, in his classical story of *La Dame aux Camélias,* was "one of those who have depicted with unerring accuracy the strange variableness of mood, the eager feverishness of life, the rapid alternations of hope and despondency which are such familiar features of the disease" (Myers, 1927).

One of the great poets of modern times, Schiller, who himself suffered from tuberculosis, described the disease in his *William Tell:*

> With noiseless tread death comes on man,
>   No plea, no prayer delivers him;
> From midst of busy life's unfinished plan,
>   With sudden hand, it severs him;
> And ready or not ready,—no delay,
>   Forth to his Judge's bar he must away!

The disease has formed the basis for many other great works of literature by dramatists, novelists, and poets. Eugene O'Neill, in his play *The Straw* describes on the basis of his own experiences the reactions of the eighteen-year-old Eileen Carmody to her environment in a small New England sanatarium. Thomas Mann, in his novel *The Magic Mountain,* depicts with great literary charm the reactions of young Hans Castorp, visitor from northern Germany to his cousin, recuperating in a sanatorium in Davos, Switzerland.

Finally, one may quote from a recent novel by Remarque, *Heaven Has No Favorites*. Although published in 1961, this appears to be based upon experiences in a Swiss sanatorium in the middle 1940's: "The coughing had stopped. Lillian Dunkerque lay back exhausted. She had offered her morning sacrifice; the day was paid for, and last night as well. She waited for the nurse to come for her. It was time for the weekly fluoroscope. She knew the routine to the point of nausea; nevertheless, it made her nervous every time . . . She hated the intimacy of the x-ray room. She hated standing there naked to the waist, feeling the assistant doctor's eyes on her. She did not mind the Dalai Lama (the chief surgeon). To him, she was a case; to the assistant, she was a woman. It did not bother her so much that she was naked; it bothered her that she was more than naked when she stepped behind the screen. Then, she was naked beneath her skin, naked to the bones and to her moving and pulsating organs. To the eyeglasses twinkling in the reddish dusk, she was more naked that she had ever seen herself, or ever could . . . She died six weeks later, on a bright summer afternoon so still that the landscape seemed to be holding its breath. She died quickly and surprisingly and alone . . . Her face was distorted; she had suffocated during a hemorrhage, and her hands were close to her throat; but a short while afterward her features smoothed and her face became more beautiful . . . She was believed [to have] been happy, insofar as any human being can ever be called happy."

PREVALENCE OF TUBERCULOSIS

According to Krause (1928), the course of tuberculosis in history took place somewhat as follows: "The disease was widely prevalent in the cities of the Greek and Roman civilizations. With the disruption of the Western Empire in the fifth and sixth centuries decentralization of population took place in Western Europe, with migration from the cities and the development of separate and widely dispersed agricultural economies. Communities were small and transportation poor. The conditions for rapid transfer of infection were lacking, and probably the total incidence of tuberculosis decreased greatly. In the craftsman industrialization of the Middle Ages centralization of population recurred, with the building of many small walled towns, intensely crowded. Presumably under such conditions tuberculosis increased rapidly. From then until the present century it must have been endemic in all civilized countries, with high mortality."

In London in 1655, tuberculosis caused more than 20 per cent of all deaths (Brownlee, 1918). The relative mortality declined to 13 per cent in 1715, but it rose again steadily, until in 1801 tuberculosis was the cause of 30 per cent of all deaths that year.

Long (1940a) states that "The maximum morbidity from tuberculosis in England apparently occurred about 1780, when the recorded mortality rate for 'consumption' was 1,120 per 100,000 population. The peak was coincident with the early development of what in later years was designated the 'industrial revolution', and it appears that the history of tuberculosis has repeated itself with respect to the industrial revolution in all other countries. The first effect has always been a rise, soon followed by a fall in the tuberculosis rate. In London the figure dropped to 716 in the decade 1801-1810 and to an average of 567 for the years 1831-35. The improvement is credited to amelioration of the bad working conditions which were brought about by the early rapid industrialization, and to other development of social measures for the relief of poverty and distress."

In 1815 Thomas Young in his book *A Practical and Historical Treatise on Consumptive Diseases* (cited by Dubos and Dubos) stated that not one patient in a thousand recovered from consumption without medical aid, and that the best possible care could save, at most, the life of one in a hundred.

The situation was no better in the eastern cities of the United

States, where the principal cause of death during the early part of the nineteenth century was not cholera or yellow fever, but tuberculosis. In 1850 Lemuel Shattuck stated in his *Report of the Sanitary Commission of Massachusetts* (as quoted by Clark, 1961): "Consumption, that great destroyer of human health and human life, takes first rank as an agent of death." Shattuck reported many figures that told the tragic story of the numerous deaths from tuberculosis, as compared with the total deaths (table 1).

TABLE 1.

MORTALITY FROM TUBERCULOSIS IN CERTAIN AMERICAN AND EUROPEAN CITIES EARLY IN THE NINETEENTH CENTURY

City	Period	Deaths from all causes	Deaths from tuberculosis	Total per cent of deaths
Portsmouth, N.H.	1800-11 to 1818-25	2,367	471	19.8
Providence, R.I.	1841-1845	3,032	718	23.7
New York, N.Y.	1811-1820	25,896	6,061	23.4
Philadelphia, Pa.	1811-1820	23,582	3,629	15.4
London, England	1840-1847	397,871	57,047	14.3
Paris, France	1816-1819	85,339	15,375	18.0

SOURCE OF DATA: Shattuck (1850).

In New York City in 1868 the annual tuberculosis mortality per 100,000 persons reached 533. Drolet (1923), who gives this figure, points out that in 1921 the rate had dropped to 103.

Robert Koch wrote in 1882: "If the number of victims which a disease claims is the measure of its significance, then all diseases, particularly the most dreaded infectious diseases, such as bubonic plague, Asiatic cholera, etc., must rank far behind tuberculosis. Statistics teach that one-seventh of all human beings die of tuberculosis, and that, if one considers only the productive middle-age groups, tuberculosis carries away one-third and often more of these . . ."

Hirsch (1883), reporting on the occurrence of scrofula (tuberculosis of the lymph nodes in the neck) in the nineteenth century, stated that: " . . . approximately half of the English population had the disease. In 1844, all 78 boys and 91 of the 94 girls in a workhouse in Kent were found to be suffering from it, although only a few of them had shown obvious signs before being admitted to the institution. Similarly, 53 per cent of the children in a Berlin orphanage were found to be scrofulous. In the French army, the mean rate of sickness from scrofula was of the order of 17 per 1,000 in 1850. It is among the children brought from the East and West Indies and from the South Sea Islands to be educated in England that tuberculosis of the joints and glands seems to have caused the greatest ravages."

In his treatise on tuberculosis Webb (1936) states: " . . . The most marked and prolonged rise in the English tuberculosis death rates, that of the eighteenth and early nineteenth centuries, corresponds in time with the beginning of the industrial revolution and the great movement of population from the countryside to the cities [and] illustrates one of the cardinal points in the epidemiology of tuberculosis, the terrible deadliness of the disease to newly exposed populations. We can hardly give adequate emphasis to the historical importance of this fact. Tuberculosis, so virulent and fatal in fresh soil, has been a major ally of the urbanized European-American white man in the conquest of new territory, the demoralization and destruction of aboriginal peoples . . . In Vienna, the tuberculosis mortality rate rose from approximately 500 per 100,000 in 1750 to 800 per 100,000 in 1870; since which time there has been a steady decline. In Sweden there was an increase in mortality from phthisis from 1750 to 1820 . . . [The] tuberculosis death rate for Boston, New York and Philadelphia combined reached its highest peak, namely 475 per 100,000, from 1812 to 1830; the rate remained from 300 to 400 for fifty years, and has since steadily declined, reaching about 80 in 1930, and even lower since then."

Baldwin (1913) analyzed the factors contributing to the mortality from tuberculosis. These were "density of population, occupation, social condition, race, color, and sex." Occupation and social conditions appeared to have been most important. "Stonecutters, cigar makers, and plasterers head the list, with about half the deaths in these occupations; farmers and persons under the best social conditions have less than one-eighth due to tuberculosis. The greatest mortality from tuberculosis is between the fifteenth and forty-fourth year of life, when it causes one-third of all the deaths occurring during that period. The months of greatest mortality are March, April, and May, when other respiratory diseases are at their height."

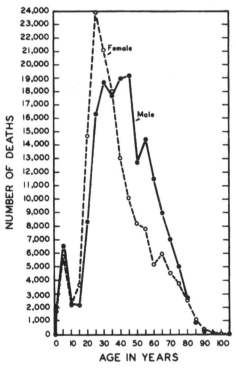

FIG. 2. Tuberculosis mortality in the United States, 1929-1932. (AFTER PERLA AND MARMORSTON, 1941.)

The mortality from tuberculosis in the years from 1929 to 1932, in the United States, is illustrated in figure 2. The Jewish people were found to show a high resistance to the effect of tuberculous infection. This phenomenon has received considerable attention. These data were based on mortality statistics collected in certain European countries where religious affiliations were recorded in the census reports. According to Perla and Marmorston (1941), "Lombroso states that in Italy the Jews of Verona have a lower mortality from tuberculosis than the Catholics of the city. In London, the Jews of Whitechapel district have only about one-half as many deaths due to this infection as the general population of the city." On the other hand, the

TABLE 2.

TUBERCULOSIS MORBIDITY AND MORTALITY BEFORE AND AFTER
INTRODUCTION OF CHEMOTHERAPY IN THE UNITED STATES AND FRANCE

Year	New cases	Tuberculosis deaths	Death rate[a]
	UNITED STATES		
1930	124,940	88,010	71.0
1947[b]	134,946	48,064	33.0
1950	121,742	33,959	22.0
1953[c]	106,925	19,393	12.0
1959	......	11,429	6.5
Decline from:			
1947 to 1953	21%	60%	64%
1947 to 1959	....	76%	80%
	FRANCE		
1930	.....	65,803	158.0
1947[b]	67,877	31,707	79.0
1950	62,107	24,282	58.0
1953[c]	60,074	15,687	37.0
1959	.....	10,487	23.4
Decline from:			
1947 to 1953	11%	51%	53%
1947 to 1959	....	67%	70%

[a] Per 100,000 population.
[b] After streptomycin was introduced into clinical use.
[c] After INH was introduced into clinical use.
SOURCE OF DATA: Drolet and Lowell (1955, 1962).

American Indians, when confined to barracks or permanent reservation quarters, show a marked rise in the tuberculosis death rate which far exceeded that of either negro or white persons: "Many of the children of Sioux prisoners of war died of tuberculous peritonitis, and scrofula was common among them, although tuberculosis of the lymph glands is not seen in children of wild tribes. Though the Indians have long been in contact with white persons, their contact has not been as intimate as that of the negro. The tepee life does not favor contact spread of tuberculosis as much as the cabin life does. The hygienic conditions are apt to be better if the Indian wanders from place to place, and his contact with others is less."

Wolff (1938) noted the close relation between low economic status and morbidity as well as mortality from tuberculosis. Stern (1941) brought out the fact that in 1935/36 people on relief had a tuberculosis rate eight times as high as those earning $5,000 a year or more, although the population ratio of the lower-income to the higher-income group was 4 to 1. A glimpse into the

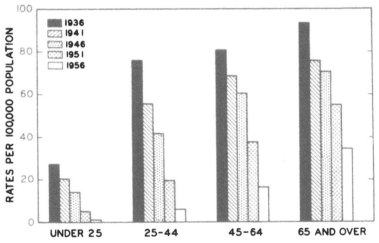

Fig. 3. Tuberculosis death rates by broad age groups, United States, 1936-1956. (AFTER *Tuberculosis Chart Series*, 1958.)

mortality of tuberculosis in the United States by broad age groups, covering the period from 1936 through 1956, can be obtained from the data presented in figure 3. The mortality from tuberculosis in the early years of the nineteenth century for certain American and European cities is set forth in table 1. Morbidity as compared with mortality in the United States and France from 1939 through 1959, as affected by chemotherapy, is shown in table 2.

According to Huber, writing at the beginning of this century (1906), tuberculosis was responsible for the yearly deaths of more than 150,000 in the United States and the average age at death was thirty-five years. "At this age the normal after-lifetime is about 32 years, so that the real loss of life covered, measured in time, is represented by 4,800,000 years per annum. . . . The mortality from tuberculosis is, therefore, a problem compared with which all other social problems of a medical character sink into insignificance, and it is safe to say that the possible prevention of a large portion of the mortality from this disease is justly deserving of the solicitude, the active personal interest, and liberal pecuniary support of all who have the real welfare of the people of this nation at heart."

Huber added: "If a case of bubonic plague, or of yellow fever, or Asiatic cholera, were to develop in New York City there would be newspaper scarehead extras, and the greatest alarm would be felt; and business and traffic to the city from the surrounding country would certainly be demoralized. Yet, as we have noted, but little attention is given to consumption, a disease much more deadly than all of these put together, and much more inimical to human happiness. The terrible Black Death lived one year in London; it killed fifty thousand. Consumption kills in the world at least 5,000,000 yearly; in England and Wales alone more than sixty thousand a year. And of all deaths in the United Kingdom between the ages of 25 and 35 nearly one-half are due to this disease."

One must record, however, a highly encouraging fact especially noticeable from the beginning of this century—a gradual reduction in the mortality from tuberculosis. Although the causes of the reduction would be difficult to prove, most students of this disease believe that better living conditions, better nutrition, better medical care, and the development of specialized tuberculosis institutions have been the major factors.

Baldwin observed in 1913: "One of the most encouraging and instructive facts is the remarkable decrease in the death rate in civilized countries during the last fifteen years. It is at once a confirmation of the value of hygienic measures and one of the obvious benefits of improved conditions of life for the laboring classes in large cities. From being the greatest scourge during the nineteenth century, when Good thought it no exaggeration to estimate that one-fourth of the population of Europe died of consumption, it is becoming so greatly lessened that in twenty-five years [between 1885 and 1911], the rate for Prussia fell one-half, or from 3.1 to 1.53 per 1,000 living. The same is true of England, where a decrease of one-half has taken place among males and two-thirds among females in forty years. In the United States there was a decrease of 54.9 per 100,000 living between 1890 and 1900, the rates being 254.4 and 109.5 respectively. In the decade from 1900 to 1910 the rate for the United States Registration Area has declined from 196.9 to 160.3, an estimated saving of 27,000 lives. This decrease is principally in the large cities, New York being the most conspicuous by a lessening of 40 per cent in sixteen years."

Another three decades elapsed, however, before the advent of chemotherapy and the dawn of a new optimism. Even then, because of the many disappointing experiences with earlier so-called cures, the medical profession maintained a cautious attitude. They doubted that the remarkable results being reported would bring about any sudden improvement in tuberculosis control. As an example of this fairly general attitude, we may quote from Hart. In 1946 he wrote:

"Should a real measure of success be reached with one or a number of chemotherapeutic agents this is not likely by itself to lead to the eradication of tuberculosis. During the past decade much else has been added to the understanding of this disease and to the ability to control it. Fuller appreciation of the importance of social factors in its incidence and spread, and of the value of social assistance (both financial and re-abling) in consolidating treatment; greater emphasis on the factor of household contact and on earlier diagnosis by means of mass radiography; and the coming to maturity of the various methods of collapse therapy and other surgical procedures: these are among the features of this period in the more favourably placed countries, such as Britain. It is improbable that success in chemotherapy will supersede all the tried and trusted methods of control acquired through the years which are applied to the individual patient, to his family, and to the community. Thus, rest can be expected to remain the foundation of treatment, and surgical methods to be needed in certain types of cases; the conditions of life to which the patient returns will surely affect critically the ultimate results of even the most spectacular chemotherapy; and the state of housing and nutrition of the people generally may be expected to continue to influence the secular trend of tuberculosis mortality and incidence. There are in the world perhaps between 10 and 20 million sufferers from active tuberculosis. In order to reduce this inroad on world health we shall probably need most of the reasonable measures—social and economic, preventive and therapeutic—that we possess now or that we can acquire in the future. The attack will remain multiple; the tactics will change."

It appears now, a little less than two decades later, that such prevailing pessimism concerning a possible major role by chemotherapy in the control of tuberculosis was hardly justified. Other antibiotics had already demonstrated their effectiveness in the treatment of various infectious diseases. Streptomycin was barely mentioned despite the fact that its discovery and its revolutionary efficiency in the treatment of tuberculosis in man had been announced to the world two years previously.

The beginning impact of tuberculosis chemotherapy was still being ignored three years later. In a treatise entitled *Tuberculosis: A Global Study in Social Pathology,* published in 1949, Mc-

Dougall followed the traditional pattern: "Infection, morbidity and mortality rates from this disease are guides, not merely to its prevalence, but to the entire social fabric of any community. There is practically no aspect of human relationship which does not affect, to a greater or lesser degree, the incidence of tuberculosis, and it has been well said that the best barometer of any country's progress in civilization, health and welfare, is to be found in the annual returns of the number of its people who die from this disease . . . Low national budgets have their counterpart in the poverty of the individual citizens; a poor national economy leads to a higher percentage of people being unable to purchase essential food supplies, and an anti-tuberculosis scheme which fails to meet the basic needs of the tuberculous population makes it difficult, if not impossible, for individual patients to receive the treatment they require."

One may leave the historical background of the great white plague on this pessimistic note. Any reader who is interested in various aspects of the history of this disease will find further information in numerous treatises on the subject, such as those of Long (1928) and Piery and Roshem (1931).

There are, however, certain other phases of tuberculosis that should be given consideration in any historical analysis of its background and its effect on human lives. It was widely believed that the disease exerted a profound influence on literature and the arts. This facet of tuberculosis lore and the early attempts at diagnosis and treatment are the subjects of the next two chapters.

«2»

Tuberculosis in Literature and in the Arts

No other disease has played such an important role in the literary and artistic worlds as tuberculosis. Many great literary and artistic figures have died from tuberculosis at an early age. This has made so profound an impression as to lead some historians to suggest that there is a definite connection between tuberculosis and genius; the belief has been expressed that mental activity is greatly stimulated by certain substances, designated as toxins, produced by the infecting organism.

As far back as the seventeenth century, Richard Morton (cited in Flick, 1925) noted the fact that consumption has a particularly striking effect upon "young men that are in the flower of their age, when the heat of the blood is yet brisk, and therefore more disposed to a feverish fermentation . . ." He also directed attention to the association of hypochondriasis and hysteria with pulmonary tuberculosis. These morbid psychic states frequently were believed to be the causative agents of the disease. It is now believed by some that they are a consequence or a direct result of the tuberculous toxemia. In 1926 Munro wrote: "There is no disease in which the mental and moral characteristics of the patient are so profoundly modified, and with which psycho-neuroses are so constantly associated, as chronic pulmonary tuberculosis. Here also the toxins of the tubercle bacillus which play so prominent a part in the evolution of the ordinary symptoms of the disease may be invoked as responsible in whole or in part for the kaleidoscopic series of mental phenomena to be met with at all stages of the disease . . . But if the profession has paid but little heed to the psychological problems so often presented by those who suffer from pulmonary tuberculosis, they have not escaped the notice of certain writers of fiction, who have portrayed the morbid psychic states which are in many cases so characteristic of the victims of tuberculosis, with remarkable fidelity . . . Did these men and women of genius accomplish their work in defiance of the deadly malady which sapped their physical powers, and in too many cases cut them off before the flower of genius had fully unfolded? Or, are we to believe, as is suggested by Jacobson and others, that the disease itself played some part in the efflorescence of their intellectual powers, and the toxin of tuberculosis actually stimulated and quickened their creative capacity?"

In his introduction to a book by Myers, *Fighters of Fate* (1927), Mayo says: "Many poisons have a stimulating effect on mental processes . . . Intoxication in the usual sense often accompanies intoxication in the medical sense. Almost any infection may produce delirium, which is mental stimulation carried to its extreme . . . [Thus] there should be a point at which the effect would be no more than exaltation, and those microbic poisons which produce only a slow effect might be expected to stop at the stage of mild stimulation. We know that victims of chronic tuberculosis have learned the significance of the feeling of unusual vitality and vigor that often precedes increase in cough, slight fever and another bout with the enemy. Man is not the only creature whom nature has cunningly equipped for the struggle of life; the little speck of living matter known as the bacillus of tuberculosis paves the way for its destructive action by stimulating its host to over-activity."

Any list of eminent figures who suffered or perished from tuberculosis would include some of the greatest men and women among the poets and novelists, essayists and philosophers, musicians and painters, statesmen and physicians. This disease is said to have a peculiar effect upon the sufferer, being distinct in this respect from other diseases, infectious or noninfectious in nature. According to Fishberg (1932): "As an exquisitely chronic disease, phthisis is accompanied by many morbid manifestations of the nervous system; in fact, nearly every symptom of the disease is often influenced by the effects of the tuberculous toxins on the nervous system. The neurotic phenomena may make their appearance immediately at the outset, in some they precede the actual onset of phthisis, while most confirmed consumptives have a psychology peculiarly their own, and show symptoms of nervous aberration which cannot escape the vigilance of the observant physician."

Out of the immense gallery of eminent personages thus afflicted, certain representatives may be picked from the gallery presented by Ebstein, is his treatise *Tuberkulose als Schicksal* (1932): Calvin, the great religious reformer; Cicero, the great political leader; Cardinal Richelieu; Cecil Rhodes; the infamous Hitler; the writers Molière, Goethe, Schiller, Gorki, Dostoevsky, Chekhov, Sir Walter Scott, D. H. Lawrence, Eugene O'Neill, Stevenson, Poe, Kafka, Laurence Sterne, and Rousseau; the philosophers Spinoza and Voltaire, Samuel Johnson, F. A. Bebel, and L. Börne; the actresses E. Rachel and E. Duse; the artists Watteau, Modigliani, Gauguin, and Beardsley; the musical composers and performers Chopin, Paganini, Purcell, and C. M. von Weber; the physicians and scientists R. T. H. Laënnec, E. Trudeau, E. R. Long, Paul Ehrlich, and many others.

The philosopher Baruch Spinoza ("the blessed Spinoza"), living in the seventeenth century, showed the salient features of a temperament usually associated with the *spes phthisica* (fig. 4). He died a victim of tuberculosis in 1677. Spinoza is said to have had the greatest influence of any modern philosopher, except perhaps Kant, upon philosophers, scientists, and poets (such as

FIG. 4. Baruch Spinoza.

Goethe, Wordsworth, and Shelley), who, according to Jacobson (1926), "have gone to him for inspiration, and the essence of his thought has been in large part appropriated in the poetic pantheism of modern interpretations of nature"

Commenting upon Havelock Ellis's *A Study of British Genius,* which listed about forty British personages, each considered a genius and all of whom suffered from tuberculosis, Myers said: "What is true of British genius is no doubt true for the most part of other nations. Tuberculosis does not produce genius, but the life of physical inactivity which the tuberculous patient is frequently compelled to live may give him an opportunity to discover or to develop his native power. Such is the case of Eugene O'Neill. Tuberculosis is accredited with causing a mental exaltation and increased excitability, during which great visions and plans for their realization come to the patient. Chopin is said to have been motivated in the composition of some of his masterpieces by such a condition. In some measure this may be true, but the greatest opportunities for increased mental activity as a result of the decreased physical capacity account for most of the relationship between tuberculosis and genius."

Moorman (1940) quotes a British physician's opinion that "Undoubtedly, a sort of restless agitation was produced in certain phases of pulmonary tuberculosis. There was also a feeling of apprehension lest life should be shortened. Both the apprehension and the restlessness might act as a stimulus to production and particularly to the production of authors, who could pursue their calling without much bodily exertion." According to Dubos and Dubos (1952), "Tuberculosis, being then so prevalent, may have contributed to the atmosphere of gloom that made possible the success of the 'graveyard school' of poetry and the development of the romantic mood. Melancholy meditations over the death of a youth or a maiden, tombs, abandoned ruins, and weeping willows become popular themes over much of Europe around 1750, as if some new circumstance had made more obvious the ephemeral character of human life. Instead of singing of the healthy joys of love, poets cultivated the refined sadness evoked by the thought that the beloved might soon depart."

The fates of some of the great musicians who suffered from tuberculosis can best be illustrated by those of Paganini and Chopin, who both went south in search of convalescence. According to Myers, "Paganini's health had so completely broken down that his landlord made a diagnosis of consumption. He was living in Naples and, because in that place there was so strong a belief that consumption was contagious, the landlord threw Paganini and his equipment into the street. Ciandelli, a friend, was passing by and saw the episode. He was not convinced of the contagiousness of consumption and certainly did not approve of this form of open air treatment; consequently he gave the landlord a sound beating. This being well done, he found good quarters and excellent care elsewhere for the tuberculous violinist."

Chopin was not strong from childhood (fig. 5). He was spoken of as a little, frail, delicate elf of a boy, and it was said that the attention of his family was concentrated upon his health. Myers points out that "It is well known fact that his family had a history of tuberculosis. In 1826, the same year in which Laënnec died of tuberculosis, Chopin, then only sixteen years old, began to show signs of breaking. His sister, Emilia, was at Reinerz suffering from far advanced tuberculosis. Chopin was sent to stay with her but she died at the age of fourteen while Chopin apparently recovered his usual health."

Chopin later went to the island of Mallorca, where he hoped to recuperate from his disease. George Sand wrote of him in 1839: "Poor Chopin, who has had a cough since he left Paris, became worse: we sent for a doctor—two doctors—three doctors —each more stupid than the other, who started to spread the news in the island that the sick man was consumptive in the last stage. As a result there was great alarm. Phthisis is rare in these climates, and is looked upon as contagious. We were regarded as plague-infested; and, furthermore, as heathens, as we did not go to the mass. The owner of the little house which we had rented turned us out brutally, and wished to bring an action against us to compel us to limewash his house, which he said we had infected. The law of the island plucked us like chickens." At Barcelona, as they were leaving the inn, the landlord demanded to be paid for the bed on which Chopin had slept, on the pretext that it was infested, and that the police regulations required that it should be burned.

FIG. 5. Frédéric Chopin, by Delacroix.

Chopin described his own situation: "I have been sick as a dog the last two weeks; I caught cold in spite of 18 degrees C. of heat, roses, oranges, palms, figs and three most famous doctors of the island. One sniffed at what I spat up, the second tapped where I spat it from, the third poked about and listened how I spat it. One said I had died, the second that I am dying, the third that I shall die. . . . I can scarcely keep them from bleeding me . . . All this has affected the Preludes and God knows when you will get them."

One of the most famous literary families afflicted with tuberculosis was that of the Brontës. The three of the sisters who reached adult life displayed extraordinary intellectual faculties. According to Jacobson: "The girls seem to have been of the precocious type which we associate with the physical delicacy that so often, in the young, denotes a tendency to tuberculosis. All of the Brontë sisters and the brother, Branwell, died of tuberculosis, although in Charlotte's case childbirth was an associated factor. Elizabeth and Maria died when they were about twelve years old, Anne, Emily and Branwell near their thirtieth years, and Charlotte when she was thirty-eight years old." Bentley (1948) believed that "The Brontës' Irish ancestry no doubt gave them a predisposition to tuberculosis," but also speculated about the deaths of the brother and two sisters and why at this particular moment the disease should gallop them thus headlong into the grave. There were of course physical reasons. Branwell died of the effect on his lungs of chronic drink and drugs, and as the long years of agony he had caused his sisters closed thus, no doubt they experienced a serious reaction, almost a collapse, after the strain. Emily perhaps caught cold at Branwell's funeral . . . The girls lived in close quarters, sharing bedrooms and always together; modern precautions against infection were unknown. But when all that has been said, one cannot but feel that these two deaths within six months, of young women thirty and twenty-nine respectively (i.e., past the most dangerous tubercular age), must have had some striking psychological cause. Anne and Emily were inseparable friends in life; it was not altogether surprising that the milder and weaker and younger girl, who had always been delicate, should make haste to follow the stronger of the pair to the grave."

In Robert Louis Stevenson we find another striking illustration of the *spes phthisica*. According to Jacobson, "The immediate cause of his death was cerebral apoplexy, but for years he had suffered from tuberculosis and had done his best work during this period of pathology. The occasional notes of despondency in Stevenson's letters from Vailima seem to have coincided with periods of temporary improvement and the literary work that he did at such times was not great. When the disease again gains the upper hand we have *Weir of Hermiston,* which, had he lived to finish it, would perhaps have been his greatest work. Finally, the *Letters* tell us that he was buoyant and happy, when very near the grave."

The medical history of the poet Keats is described by Hale-White (1938) as follows: "He was a healthy young man. There is no reason to believe that he was born with any special predisposition to consumption. His mother who died in 1810 may have had it, but he did not catch it from her, for he was in excellent health for many years after her death. In August 1818 he had a sore throat, which occasionally troubled him till December 1819 . . . He caught consumption almost certainly from Tom (his brother), with whom he lived for months before it killed Tom in December 1818. From this date until February 1820 Keats occasionally complained of not feeling well—he did not bathe for this reason. Whether this was owing to tubercle bacilli in his lungs, or to some other cause, we do not know; the tubercular disease of his lungs first showed itself on the 3rd of February 1820 when, fourteen months after Tom's death, he coughed up blood. The disease then progressed in Keats in the usual way until it killed him in February 1821. To doctors and to present-day readers the treatment of Keats seems horrible, and it was horrible, but it is difficult for us to think of consumption as it appeared to doctors in 1820, for they were unaware of the fundamental fact, not discovered for nearly a century, that it is due to the tubercle bacillus. They saw no harm in Keats living in close contact with his brother Tom who was coughing up these bacilli into the air breathed by both brothers, and hence Keats inhaled them into his lungs where they produced no definite symptoms until he coughed up blood."

Keats wrote to Shelley: "There is no doubt that an English

winter would put an end to me, and do so in a lingering, hateful manner. Therefore, I must either voyage or journey to Italy, as a soldier marches up to a battery. My nerves at present are the worst part of me, yet they feel soothed that, come what extreme may, I shall not be destined to remain in one spot long enough to take a hatred of any four particular bedposts. I am glad you take any pleasure in my poor poem, which I would willingly take the trouble to unwrite, if possible, did I care so much as I have done about reputation . . ." C. A. Brown (1937) intimates that Keats was doomed to die of tuberculosis in any case: ". . . he probably would have died from it, as did his mother, brothers, and four nieces and nephews, for the debilitating treatment of the disease was exactly counter to modern medical practice. But his mental condition, irritated almost beyond endurance, must have hastened the progress of his illness. The critical attacks were the starting-point for his melancholia; they therefore must be regarded . . . as important contributory causes of his death." Shelley himself suffered from chronic tuberculous pleurisy.

Sidney Lanier, the lyric poet, is considered a perfect type of the tuberculous genius. Jacobson says: "He produced nothing of any consequence until after he became afflicted in 1868 or perhaps as early as 1865. His work improved as his malady advanced. Making all allowance for natural improvement in technique and for increased intellectual breadth, we cannot summarily dismiss consideration of the phthisical element. It stirred his soul into expression in the beginning and as time passed on became more and more excitative. *Sunrise,* considered the 'culminating poem,' when he was rapidly approaching the end, was composed at a time when he was running a temperature of 104° and was unable to lift his hand."

Various other American poets, and also philosophers, dramatists, and novelists, have been afflicted with tuberculosis. Some managed to adjust themselves and survived; others succumbed at an early age.

Thoreau, for example, came from a consumptive family. His

grandfather had died of the disease in 1801. His brother John showed signs of tuberculosis, although he died from lockjaw contracted in an accident. His sister Helen had been consumptive from girlhood and died in 1849. His father died in 1859, apparently from consumption. Canby (1939) analyzed Thoreau's condition as follows: "He had his narrow rattan day-bed, made by himself, brought down to the parlor for sociability, and as long as he could do so insisted upon eating with the family at the table. His voice sank to a whisper, he had an incessant cough, but his serenity was never disturbed. Sam Staples, once his jailer, later his rodman on surveying jobs, came to see him. 'Never spent an hour with more satisfaction,' he told Emerson. 'Never saw a man dying with so much pleasure and peace.' [When Channing spoke to Thoreau] of the approaching solitude, he murmured, 'It is better some things should end.' And so, quietly, at nine in the morning of May 6, 1862, having been lifted upright on his couch, he died imperceptibly with neither apparent struggle nor pain."

Ralph Waldo Emerson suffered from a chronic form of tuberculosis. He too came from a consumptive family; all four Emerson boys developed tuberculosis. He suffered from "Oppressions and pangs chiefly by night [and] a certain stricture on the right side of the chest, which always makes itself felt when the air is cold or damp." According to Moorman, who quotes Emerson's own words, he possessed a dual personality: " . . . one part 'toiled, compared, contrived, added, argued'; the other 'never reasoned, never proved; it simply perceived; it was vision; it was the highest faculty.' " Emerson went south in November, 1826, to escape the winter, but "the following May we find him writing: 'I am still saddled with the demon stricture (pleurisy), and perhaps he will ride me to death. I have not lost my courage, or the possession of my thoughts . . . ' Approximately a year later he said: 'It is a long battle this of mine betwixt life and death, and it is wholly uncertain to whom the game belongs . . . ' " He managed, however, to overcome the acute form of his disease and lived to the age of seventy-nine.

Among the more recent writers who suffered from tuberculosis, it is sufficient to mention Eugene O'Neill, who, while working as a newspaperman in Connecticut, became ill of this disease. Of him, Myers says:

"He was twenty-five years of age when he entered Gaylord Farm Sanatorium. Dr. David Lymann, one of America's most favorably known tuberculosis workers, was superintendent . . . Up to this time O'Neill had pursued no definite plan in life, but his stay in the sanatorium completely changed his outlook . . . Even though O'Neill has [since] accomplished a great deal in the literary field he has kept a careful watch over his health."

Of the great dramatist and poet Schiller, Jacobson says: ". . . the disease served in some way to increase activity and fan his intellect into a keener flame [and] he produced his finest work when his health was gone." Carlyle describes the final effect of tuberculosis upon Schiller as follows: "The poet and the sage was soon to lie low; but his friends were spared the farther pain of seeing him depart in madness. The fiery canopy of physical suffering, which had bewildered and blinded his thinking faculties, was drawn aside; and the spirit of Schiller looked forth in its wonted serenity, once again before it passed away forever. After noon his delirium abated; about four o'clock he fell into a soft sleep, from which he ere long awoke in full possession of his senses. Restored to consciousness in that hour . . . Schiller did not faint or fail in this his last and sharpest trial. Feeling that his end was come, he addressed himself to meet it as became him; not with affected carelessness or superstitious fear, but with the quiet unpretending manliness which had marked the tenor of his life."

Dostoevsky was another victim of tuberculosis, as Myers points out: "Although for the greater part of his life Dostoevsky was in poverty and suffering from two chronic maladies—tuberculosis and epilepsy, yet critics say of him that he is not only the greatest of all Russian novelists, but also one of the greatest writers the world has ever produced." Among other eminent Russian literary men who died early from tuberculosis should be mentioned the poet S. I. Nadson, who died at the age of twenty-four, and the critic and philosopher V. G. Bielinski, who died at the age of thirty-seven.

Myers tells us that the English artist Aubrey Beardsley continued his work despite the presence of advanced tuberculosis: "His latest drawings, for the illustration of *Mademoiselle de Maupin* and for *Volpone,* some of the latter produced only four or five weeks before his death, show no falling off in vigour and inventiveness, but rather a fresh development in technique. It is pitiful to think of him lying prostrate, his active brain teeming with ideas which his poor, wasted hand was unable to express; to think of him asking for pencil and paper, and, after a few unavailing efforts to commit his ideas to definite shape, having to let the pencil drop from his enfeebled fingers."

Anatole France describes, in his book *The Latin Genius,* the suffering of the French poet Paul Scarron: "Came autumn, and with autumn rain, gloom, and the melancholy sighing of the wind in the trees of Saint-Gervais. What he suffered is incredible. At last came a hiccough, a fatal sign. 'What a satire will I write,' said he, 'on this hiccough, if I get over it.'" The poet composed his own epitaph:

> He who here now sleeps beneath
> Roused more pity than ill will,
> Living suffered many a death
> Ere life left him still.
> Pass, O friend, with footfall light,
> Lest he wake—'t would be too bad
> Should you break the first good night
> Scarron ever had.

Among the statesmen who have suffered from tuberculosis, the name of Cecil Rhodes is usually mentioned. In speaking of him and his companions in South Africa, Millin (1933) writes: "Strange to think of these three men, these three sickly bachelors, all born in the same year [1853], an Englishman, a Scot, and a German Jew, making this great, untamed country the work of their lives. Rhodes had tubercular lungs and an aneurism of the aorta. Jameson had tubercular lungs, haemorrhoids, and gallstones. Beit had dangerously unquiet nerves. No one would ever have chosen them to be Rhodes Scholars. They had not been leaders or sportsmen at school. They were leaders now . . . Students they never became. Sportsmen they never became . . . The thing one fails to remember about Rhodes—so vehemently he lived—is that for a long time death was his daily companion . . . the real, immediate, frightening death that grimaces from the scaffold at a man condemned. Rhodes had come out to South Africa because death was before him. He had fled back to South Africa from Oxford because its breath was in his very face. 'You the same Rhodes, sir?' the doctor said who had once written him down as tubercular beyond recovery. 'Impossible! According to my books you have been a corpse these ten years.'"

This story of the great men who suffered and died from tuberculosis would not be complete without mention of the eminent physicians who were afflicted with this disease. In modern times, Laënnec, Trudeau, and E. R. Long, who suffered and survived, made great contributions to the final conquest of the disease.

In speaking of Laënnec, Myers says: ". . . bleeding, which had been practiced by the ancients in treating many diseases, was still employed in the treatment of tuberculosis. He had much blood drawn from his veins. One can scarcely imagine any thing more harmful than the taking of the life-blood of which he already had too little and for which he stood in so great need. From time to time his attacks grew more severe and he became weaker and weaker. He came to the conclusion that he was suffering from 'galloping consumption.' His long study of tuberculosis and his resulting written work on the subject justify one in saying that no one had lived before him whose knowledge of tuberculosis could approach his. In consequence, he was well aware of what was going on in his own case and he understood clearly, out of his own observation, that the end was not far distant."

In analyzing the relation between great mental activities and tuberculosis, Webb (1936) stated: "It is noteworthy in studying the literature of tuberculosis that so many physicians became interested in the subject either because they were victims themselves or the disease had attacked their families. Thomas Willis, whose name is so well known to all students of anatomy and to otologists, lost his wife, a daughter, and a son from tuberculosis. The latter died at Montpellier, which was a famous resort for consumptives at that time. While Willis made little fame from his prescriptions of snails and snail syrup, and inhalations of sulphur and of arsenic, yet he noted that ulceration was not present in all cases of phthisis, describing an autopsy in which there were scattered through the lungs in every part, tubercles or stones of sandy matter."

TUBERCULOSIS AND GENIUS

Moorman, reporting his observations and those of many historians of the disease, has stated that in sufferers from tuberculosis "there seems to be a strange psychological flair—a phenomenon not fully accounted for, not of established scientific lineage, yet quite evident to the student of clinical tuberculosis. Everyone who deals intelligently with tuberculosis individuals knows how patiently they bear their lengthening burdens; how courageous they are, often in the fact of insurmountable obstacles; how optimistic they may be even when life is literally being cut down by the inevitable sweep of the Great Reaper. This unusual display of courage and hopefulness has been termed *spes phthisica.*" No wonder, then, that biographers and critics "have long recognized progressive tuberculosis in geniuses as a possible factor contributing to their individual greatness," and while not contending that the disease causes genius, have believed that "it may fan into flame an otherwise dormant spark," although "advanced tuberculosis with physical prostration may inhibit creative effort."

Jacobson notes that others, in attempting to be more specific, have calculated that the "release of creative secondary personalities would seem to depend upon some sort of intoxication, with resulting paralysis of inhibitions." Jacobson adds: "This is obviously true of alcohol, as well as of the toxins of tuberculosis. They are keys wherewith some individuals can unlock the unconscious. So we find Byron 'dreaming awake' after his potations; so also the tuberculous Emerson walking in a trance-like state, during which he 'receives' and records short but complete messages . . . Dr. Robert T. Morris has pointed out the marked vulnerability of most geniuses to microbic influence. This special sensitization he believes accounts for some of the manifestations of genius; so much injury results that there is likely to be ill health, no progeny, or progeny that is feebly resistant, succumbing quickly to ordinary destructive agents. The world's history has been made by men of this neuropathic stock, he declares, lacking not only good resisting powers but deficient also in moral sense, sound judgment, and aptitude for higher control."

Jacobson says, further: "Were the present writer to give an almost sure recipe for producing the highest type of creative mind, he would postulate an initial spark of genius plus tuberculosis . . . Now it is entirely conceivable that the tuberculous by-products are capable of profoundly affecting the mechanism of creative minds in such a way as to influence markedly their creations. Indeed, they are bound to do so, for the *spes phthisica,* admittedly a result of such by-products, must necessarily affect the whole psychologic switchboard . . . Many writers have noted that tuberculosis is in some way a favoring condition of certain types of esthetic and intellectual capacity. To be sure, tuberculosis does not convert all talented persons into geniuses, nor mediocre people into talented ones. Again, tuberculosis is frequently a result, and not a cause, of literary industry, although in such instances it may prove to be an intellectual asset. The hectic afflatus of the actively tuberculous creative genius is almost incessant and he is nearly always astonishingly prolific. The inspiration of the nonphthisical genius is intermittent, his work is more deliberate, he does not burn the candle at both ends, he is normal and works sanely. The wheels are not continually in motion."

In the same vein one may quote Jeannette Marks (1925): "Genius goes back to physical fact. It is a question of sensitization of protoplasm. The foundations of the greatest cathedral of beauty ever erected by the mind of genius rest squarely upon the flesh of a man's body . . . It has been said that a man is what his microbes make him, and in nothing, it would seem, is this more true than with the man of genius . . . It should be remembered that there are types of optimism equally pathological which are due to the quick burning of disease. For example, the buoyant hopefulness created by tuberculosis. . . . [In] Shelley's *Ode to the West Wind* it is doubtful whether the flight of his song and the tumult of wind and leaves would have been so swift without the quickening which Shelley had from tuberculosis. In the case of Emily Brontë, life may have been shortened physically by consumption, but study convinces the reader that psychically in *Wuthering Heights* and in her poems power and passion were made the greater by the *spes phthisica*."

Wolff (1938) postulated that "It should never be the task of the physician, as defined by Hippocrates, to breed human types refractory to tuberculosis ignoring all other mental and physical hereditary qualities. If the 'norm' of the ideal man is not present to our senses, neither the statistician with his calculations of averages, nor the geneticist, cultivating a particular quality, nor the philosopher with his general ideas, will discover this norm. Such men as Rafael, Spinoza, Schiller, Mozart, and many others who died early from phthisis, are names we should hardly be prepared to forego in exchange for hereditary-biological qualifications and equality of breeding which might lead to the extinction of the type."

Finally, Pittfield (1930, quoted by Moorman), in speaking of Keats, said: "It is more than likely that the tuberculous poisons intoxicated and enriched the imaginations of Stevenson and Chopin and other geniuses. These poisons no doubt added much to the fervid vision of this man. I am sure that they guided his genius. A peculiar mental hyperesthesia characterizes this disease in even the most commonplace minds. There can be no doubt that Keats was hyperesthetic, acutely so."

In spite of all the speculation concerning *spes phthisica,* its reality has never been fully established. Since tuberculosis was such a common disease and since such a large proportion of the population of Europe and the Americas died from it, it was quite inevitable that many of the most talented people should have succumbed to it. A false impression was thus created that talented people were specifically afflicted by tuberculosis, and, even more significantly, that the disease-producing microbe produced a toxin which stimulated diseased persons to greater creativity. The fact that we still have among us numerous gifted people, including many who can no doubt be classified as geniuses, yet few of whom have tuberculosis, does not prove that the association between tuberculosis and accentuation of mental abilities is purely coincidental, but does show, at least, that genius can flow without a tuberculous infection. Unfortunately, the strong clinical impression concerning such an association was gained at a time when it was bound to be frequent. Although this question will never be answered in a definite manner, the weight of evidence tends to be on the side of pure coincidence rather than direct causation.

Porot, in his treatise *La psychologie des tuberculeux* (1950), states emphatically that the claim of a direct effect of the tuberculosis bacterium upon the functioning of the patient's brain cells has not been confirmed. This is contrary to the prevailing opinion in some quarters that the organism is responsible for the causation of certain mental disturbances in the afflicted individuals. The conclusion may, therefore, be reached that tuberculosis has only a negligible action upon the intellectual activity of the individual, or that rather, it causes fatigue and exerts a weakening effect upon the faculty of intellectual concentration. Such an effect was originally recognized by Laënnec, who called attention to the periods of psychic depression in tuberculous patients. This effect can accelerate and aggravate the condition, or it can slow it down and even favor healing. Among other manifestations, the psychic effect also can express itself in a defensive attitude, egocentrism, or intolerance. The various effects of tuberculosis upon the psychic state of the individual, which may be owing either to the chemical substances produced or to the organic changes that result from the special mode of living imposed by the treatment, are undeniable. This was true particularly before the advent of chemotherapy, when the sufferer had to spend much of his time in a sanatorium.

«3»

Infectious Nature of Tuberculosis
Early Attempts at Treatment

Command the children of Israel, that they put out of the camp every leper, and everyone that hath yet issue, and whosoever is defiled by the dead. NUM. 5:2

The infectious nature of tuberculosis came to be recognized in early historical times. The prevailing view varied greatly with different peoples, depending on their beliefs and state of civilization. As usual, the wrath of the gods was brought into the picture as the proper explanation for the disease. In the more advanced civilizations, the disease was considered as either contagious in nature or as hereditary; some saw in it a characteristic condition inherent in the nature of man.

The Yajur-Veda scriptures of India spoke of tuberculosis as follows, according to Webb (1936): "Consumption is a disease difficult to stop; it is endowed with great power . . . A consumptive who is evidently master of himself, who has good digestion, is not emaciated, and is at the beginning of the disease, the physician can cure . . . A consumptive who eats little, who is failing, who has diarrhea, tumefaction of the scrotum and of the abdomen, that one, the physician anxious to be a man of renown will abandon . . . The physician who wants great fame cures a man attacked by consumption."

The Hebrews appear to have been exposed to tuberculosis for

a long time, since they have developed a greater immunity to the disease than has any other race. It is possible that tuberculosis was one of the plagues that afflicted Egypt before the exodus, and it has been suggested, therefore, that the Hebrews contracted tuberculosis in Egypt. The Bible has ample references to the infectious nature of various diseases and to the destruction of the clothing and even the habitations of sufferers, but it is not clear that consumption as such was recognized, or that rigid precautions were taken against this disease such as were practiced in cases of leprosy.

Greek writings show a rather vague concept of tuberculosis. Flick (1925) says: "Several theories crop out in them. So far as one can determine, the most prevalent one was that through a disturbance of the harmony of the four interchangeable elements of the human body, namely, earth, water, fire, and air, or in other words, heat, cold, moisture, and dryness, a distillation from the head, probably from the brain, through the nostrils and trachea into the lungs, set up an ulceration which ultimately led to the destruction of the lungs and by thus cutting off the cooling air eventuated in a burning-up of the tissues of the body."

The Romans practiced climatherapeutic treatment. Phthisis was said to be cured more easily in a warm and dry climate. Castiglioni (1933) writes: "In imperial Rome, it became the custom to send people affected with pulmonary maladies to Sicily and Egypt as it was well known that they would often return cured . . . Worthy of mention is the recommendation of a long sea voyage for the tuberculosis . . . It is said that Cicero, having hereditary predisposition to phthisis, undertook long voyages to Greece, Asia and Rhodes in 80 B.C. He tells that he was gravely ill, expectorating blood; that later he lost weight, and that his neck had grown thin and flabby, and he coughed frequently. After two years he returned to Rome entirely cured . . . The old Romans—true adherents to the Greek tradition also in medicine —considered of great importance the diet and rich nutrition in pulmonary phthisis especially at its inception . . . Hydrotherapy was also recommended."

Galen (A.D. 131-200) regarded tuberculosis as incurable and contagious, and advised avoiding it. He suggested that the lungs need rest in order to recover, and that coughing injures them. For fever he prescribed residence in a cool, well-ventilated, underground room and forbade visitors, even the family. "To insure fresh milk he brought the woman or the ass into the patient's room."

The first to emphasize that tuberculosis is contagious was Girolamo Fracastoro, who in his book *De Morbis Contagiosis* (1546) devoted a chapter to contagious phthisis. "He differentiates between spontaneous phthisis following a rupture of a vessel in the chest or resulting from pleuritic effusions or some other causes—and contagious phthisis contracted directly by associating and living with people so afflicted. . . . 'The infectious element which lodges in solid bodies seems to be of a different nature [from that contained in liquid or soft parts.] It is really astonishing with what tenacity and for how long a time the particles of this virus are preserved in the solid bodies in which they become embedded, for example, clothes which a phthisical patient had worn, have communicated the disease even after two years; the same may be said of rooms, beds, floors, where a sufferer of phthisis has died. It must be assumed therefore that the seeds of contagion remained in such bodies. We must, therefore, conclude that in such *fomites* (vehicles) are left behind the seeds of contagion and it is a fact that there is an incredible analogy or selective affinity between these germs and the pulmonary tissue for they are not communicated to any other organ of the body . . . Were it possible to destroy them by the use of caustics, there would be no better remedy; but, since these substances cannot be employed without danger to this organ, one should seek to treat by way of the adjoining ones.' (Castiglioni.)"

The contagiousness of tuberculosis was well considered in the sixteenth century by Sylvius, who tried to harmonize it with his ideas of the tubercle which he had described. He listed five causes of tuberculosis as those usually accepted by the physicians of his day. In addition he mentioned contagion as a sixth cause (as quoted by Flick): ". . . the air expired by consumptives having been brought close to the mouth and nose [of other persons] is drawn in and in this way offensive and irritating emanations are continuously carried from the affected party to others especially relatives and when these are finally infected with the same poison they also fall into phthisis. And persons who have consulted their experiences and have observed accurately what usually happens to those in the environment of consumptives and to relatives during tender and younger years, will learn that this is not more than the truth; for which reason physicians quite properly advise and admonish those in the environment of consumptives and especially relatives to protect themselves against the breath of those who are afflicted; and I suspect that those are usually infected in whom there already is naturally a predisposition to phthisis because I do not find that all are equally affected . . ."

Morton (1689, as quoted by Castiglioni) had similar ideas on pulmonary phthisis: "[It is a] malady which produces consumption of the entire body; it is accompanied by fever and springs from a defective state of the lungs and from the resulting ulceration of pulmonary tissue . . . Consumption of the lungs may be either original which from the very beginning depends upon an ill-disposition and an exulceration of the lungs; or secondary and symptomatological whenever the lungs received any great injury from preceding distempers."

The contagious nature of tuberculosis was thus well recognized in the seventeenth and eighteenth centuries. This was particularly true of southern Europe. A law was passed in Spain for the prevention of consumption which covers the practical side of the problem so fully that one can infer from it a fair knowledge of the subject. In Italy, where considerable progress was made in social prophylaxis during the eighteenth century, it was even required that the disease be reported and that the belongings of its victims be destroyed. The Republic of Lucca promulgated a law in 1699 which stated: "In future the health of the human body shall not be harmed or imperiled by objects remaining after death of a person infected with the disease of phthisis." The names of patients suffering from the disease were to be reported and disinfection measures were required. The physicians were directed to perform autopsies on those dying of tuberculosis. The following edict may be cited as an illustration: "It should be the duty of those around the phthisic patient, to leave the entrance open from time to time, for egress of fresh air, and to take care that the patient does not empty his sputum except into vessels of glass or glazed earthenware, and that these utensils be frequently cleansed and boiled in lye at least twice and the same should be done with all clothes of washable wool as well with mattress and pillow ticking. [Also,] the floor of the room should be scrubbed at least twice and the walls freshly painted."

The physicians frequently recognized the danger from contact with afflicted persons. A regulation of the Department of Health of Naples emphasized that "Pulmonary consumption is of such a malignant nature in our country that even after the death of the sick person the seed of his malady remains hidden and unseen in many houses, with serious danger to those who move into them thoughtlessly; and indeed some of this seed is so penetrating that it can be communicated even without immediate contact with the infected person or thing."

Salvatore de Rensi, the historian of medicine in Italy, gave a full account of the clauses of this law, enacted on July 19, 1782, as follows (according to Flick):

"1. That the physician shall report a consumptive patient when ulceration of the lungs has been established under penalty of three hundred ducats for the first offense and banishment for ten years for repetition of it.

"2. That the authorities make an inventory of the clothing in the patient's room to be identified after his death; and if any opposition shall be made the person doing so, if he belongs to the lower class shall have three years in the galleys or in prison, and if of the nobility, three years in the castle and a penalty of three hundred ducats.

"3. That household goods not susceptible of contamination shall immediately be cleansed and that which is susceptible shall at once be burned and destroyed.

"4. That the authorities themselves shall tear out and replaster the house from cellar to garret, carry away and burn the wooden doors and windows and put in new ones.

"5. That the poor sick shall at once be removed to a hospital.

"6. That newly built houses shall not be inhabited within one year after their completion and six months after the plastering has been done and everything about the building operation has been finished.

"7. That superintendents of hospitals must keep clothing and linens for the use of consumptives in separate places."

In 1720 Benjamin Marten stated in his *A New Theory of Consumption* (cited in Cummins, 1949) that tuberculosis might be caused by "minute, living creatures" which, when they gained access to the body, produced the lesions and symptoms of the disease. Marten had knowledge of the first microscopes; this suggested to him that the principle of living "animalculae" might be applied to tuberculosis. He stated further: "[The infection] can be acquired in an hereditary manner—perhaps *in utero*—but also by contact. Casual association with an infected person is not to be feared (as in smallpox), but constant exposure is dangerous. It is especially so for those who have inherited little resistance, that is, for those who are 'hereditarily disposed' toward the disease . . . persons born of consumptive parents and such as are prone to spit up a black flegm in a morning . . . or that spit blood, though it be only accidental, through loud hollowing, singing, hard coughing, running or any violent straining . . . [These] are much more liable to a consumption than others . . . [The] minute animals or their seed . . . are for the most part either conveyed from parents to their offspring hereditarily or communicated immediately from distempered persons to sound ones who are very conversant with them . . . It may, therefore, be very likely that by habitual lying in the same bed with a consumptive patient, constantly eating and drinking with him or by very frequently conversing so nearly as to draw in part of the breath he emits from the lungs, a consumption may be caught by a sound person."

According to Dubos and Dubos (1952), the "frequency of multiple cases of pulmonary consumption in one household, and the extinction of many consumptive families, made it appear certain in the Northern countries that the disease was the outcome of a bad hereditary constitution. It was generally considered that little could be done for those born with a predisposition to phthisis beyond providing for them a climate and a way of life that would retard the inexorable course of their disease Only balmy air and sunny skies, it was thought, could arrest the destruction of lung tissue and the sapping of strength that otherwise drove the consumptive to certain death in the space of a few years."

In 1784, Natale Saliceto, a well-known Roman physician, published a book on the contagiousness of consumption. In it he used a quotation from Rousseau: "Men mutually poison each other by crowding together." The chief physician of Venice held the same opinion and induced the city officials to introduce preventive sanitary measures. The doctrine of contagiousness, however, was not universally accepted. Antonio Cocchi, a Tuscan, maintained that consumption was not contagious under any circumstances. In an effort to calm the public mind, he called to severe account those who distressed the poor by publishing statements about the contagiousness of the disease. According to Flick, "So powerful did the opposition grow that in 1809 the chief magistrate of Naples again consulted the medical faculty about the scientific facts underlying the law and the wisdom of keeping the law on the statute books. The profession was much divided upon the subject, the majority being against the doctrine of contagion and the desirability of keeping the law in force. Those who were in favor of the doctrine and for maintaining the law, however, were the foremost men in the profession."

Although real scientific evidence was still to come, the early nineteenth century ushered in an era of still greater conviction of the communicability of phthisis. This can be recognized from the work of Andral (1826, quoted by Flick), who said: "The fear of contagion of pulmonary phthisis was pushed to such a point in the centuries preceding ours that Morgagni himself confessed that he no longer dared, save very exceptionally, to do autopsies on the bodies of consumptives, for fear, said he, of contracting their disease. He kept this prejudice all his life, and in one of his letters, one reads the following sentence: Young man, keep away from the dead bodies of consumptives, I, even as an old man, keep away from them."

In 1883 Hirsch presented evidence concerning both the hereditary and the contagious nature of tuberculosis: The fact that "phthisis propagates itself in many families from generation to generation is so much a matter of daily experience, that the severest sceptic can hardly venture to deny a hereditary element in the case: even if we be unable for the present to decide whether it consists in the transmission of a specific poison, something like that of syphilis, or, in other words, whether it be heredity in the narrower sense; or whether it [depends] upon a congenital disposition towards the disease, a disposition that has to be looked for, naturally, in the organization of the respiratory system."

Drolet (1946) emphasized that "the epidemic character of tuberculosis is undoubted, though in contrast with the more acute infections it manifests itself in comparatively slow-moving cycles, sometimes through several generations or across a series of communities." He noted as general characteristics of its incidence that it is widespread at times and is common to particular localities or among certain groups; that it rises and subsides, definitely rising when it comes across virgin soil and gradually declining as resistance is evolved; and, finally, that it may "flare up again where contact has been lost."

Etiological factors in the progress of the disease have been discussed by Wolff (1938), who recognized (a) the factor of infection; (b) the congenital qualities of the individual and his hereditary constitution, and (c) the various influences of the social environment.

EARLY ATTEMPTS AT TREATMENT

Reference has already been made to the fact that sunshine and southern climate were among the early recommendations for treatment of tuberculosis. Galen, one of the first to make such specific recommendations, favored the balmy zones around Vesuvius. Even before Galen, Pliny had declared that the sun is the greatest of remedies: *Sol est remediorum maximum.*

Countless other physicians have advocated changes of climate for consumptive patients. Portal, in his *Observations on the Nature and Treatment of Phthisis Pulmonalis* (1792, cited by Webb), noted that the disease was aggravated in English patients sojourning in the south of France. Even he, however, believed that in the early stages of "scrofulous consumption" it would be advisable to seek warm sea-air.

Among the other ameliorative measures, consideration was usually given to bed rest and diet. It has been stated by Shryock (1957) that "Greek therapy seems to have been more sane than that followed throughout most of the modern era. Classical treatments may have done something for bodily resistance and, in any case, they involved no excessive bleedings or exercise."

The seventeenth-century British physician Sydenham reported that consumption was occasionally curable and that the cure in such instances was dependent wholly upon exercise. Riding on horseback every day was recommended. Anyone who had recourse to this exercise did not need to be tied down to any rules of diet, nor be debarred from any kind of solid or liquid aliment, in order to attain a cure. This recommendation was not accepted generally, and Sydenham was subjected to severe criticism for it.

Milk was a favored dietary item for centuries. Woman's milk was often favored in preference to asses' milk, but the *Code of Health of the School of Salernum* (as translated by Ordronaux, excerpted in Rapport and Wright, 1952), recommended asses' milk:

> Goats' milk and camels', as by all is known
> Relieve poor mortals in consumption thrown;
> While asses' milk is deemed far more nutritious,
> And e'en beyond all cows' or sheeps', officious.
> But should a fever in the system riot,
> Or headache, let the patient shun this diet.

Webb notes the value attributed to woman's milk by many writers. Petrus Forestus (1522-1595) advised his consumptive patients to nurse from a young woman, suggesting that they place themselves beside her so that they could suck continuously. If patients would not consent to a wet nurse, they must drink warm cow's, ass's or goat's milk. G. B. Morgagni also recommended woman's milk. Morton stated in his *Phthisiologia* in 1689: ". . . I have sometimes observed a consumptive disposition cured by giving suck, and that not only in my own dear wife, but in many other women." John Wesley (1703-1791) in his *Primitive Physick,* recommended cold bathing and breathing into a hole cut in the fresh earth, and that the patient "in the last stages [of treatment] suck an healthy woman daily." This procedure, he said, had cured his father.

Marine air was another favorite treatment. "Dans la montagne la maladie va vite," said Laënnec, who also noted the rarity of consumption at the seashore. In the belief that there was something beneficial in seaweed, he brought seaweed from Brittany to place under the beds of his hospital patients in Paris. The idea was ridiculed by others. Nevertheless, many marine resorts in Europe were recommended for the treatment of scrofula or glandular tuberculosis.

Sun radiation was also regarded favorably, Koch himself having observed that the tuberculosis organism was quickly killed by sunlight. Webb relates that "Finsen, towards the end of the nineteenth century, conducted experiments with light therapy and found that ultraviolet rays caused sunburn. He introduced light therapy for lupus. Rollier believed that patients who developed the most pigmentation of the skin from heliotherapy recovered more satisfactorily from extrapulmonary forms of tuberculosis."

In the settling of the western portions of the United States, especially the prairies and the arid regions, the migration of the "pale-faced invalids," or those who had "lung trouble" played a not insignificant role. According to Long (1940*b*), it was not only the final destination but the ride itself that benefited the health seekers, and he alludes to the old English and later also the American tradition that horseback riding and strenuous exercise were beneficial in tuberculosis. He quotes from an article published in 1880 in the *Arizona Daily Star* on "Tucson as a Sanitarium": "It is plain to be seen that such a climate must be health restoring. It is a fact worthy of note that many hundreds of people who come here in the last stages of pulmonary troubles recover in a short time so that they are enabled to engage in business. Many of the most progressive pushing and wealthy business men of Southern Arizona left their northeastern homes and came here as confirmed invalids, with the hope of dying easy. They would not die now for ten times what they are worth." Long refers also to the statement of Erna Fergusson, in her book *Our Southwest,* that "Albuquerque had two businesses, the Santa Fe Railroad and tuberculosis."

Among other historic treatments for tuberculosis, drink and food must be mentioned. According to Rapport and Wright, the Nantucket *Inquirer* in 1821 advised partaking of the following for consumption or cough: "Two quarts strong ale, one of white honey, two ounces of leaves of longwort . . . put in an earthen pot covered closely, and boiled down one half." The journal also recommended "Inhaling the smoke of rosin, while burning, or the steam of tar while boiling" as having "very powerful effect in strengthening weak and decaying lungs."

Dr. Thomas Beddoes, father of the poet, wrote in 1799 that the persons most free from consumption were precisely those that consumed animal food. Their healthfulness was undoubtedly not to be attributed to this circumstance alone, but it was to be presumed that their substantial diet had its share in determining their good condition. Children fed on a vegetable diet and deprived of animal food, he thought, were likely to be subject to scrofulous affections. Exercise was necessary to give effect to diet. He attributed the great tendency of the English people to consumption to, at least in part, a lack of physical activity: sedentary employments and restraint in school life were blamed in part for the prevalence of consumption. Beddoes held that consumption was on the increase in his day, and he deplored the seeming apathy in regard to it. He urged greater activity on the part of medical men in research on the disease and advocated the organization of a society for this purpose. "By private efforts," he wrote, "the requisite knowledge will not soon or not at all be accumulated. But it seems as if *a society for collecting information relative to dangerous pulmonary ailments* could be easily formed and its means without difficulty directed to worthy ends." Flick, on whom we have drawn, points out that Beddoes' suggestion is the first reference he has found to an effort at organization for the study of tuberculosis.

Leaving aside the various reported "cures," most nineteenth-century physicians looked upon tuberculosis as a hopeless disease. According to Shryock, those in the United States might prescribe "olive oil and a trip to Florida," but generally there prevailed among physicians an attitude of "professional nihilism," which "tempted patients to turn to the medical sects or to out-and-out quackery for aid." Benjamin Rush (1745-1813, as cited by Webb) warned students and physicians of the temptations faced by their patients: "The eagerness with which patients seize hold of everything under the name of a remedy borders on madness. For this reason a physician should if possible attend on his patient. Every town, every village, has its quacks, and prescriptions will be thrust upon him from barber shops, taverns and stables." Yet, as Shryock points out, " 'Consumption cures' were a feature of the quack advertisements of the nineteenth century. But therapeutic skepticism at least had the merit that it focused the attention of some physicians on hygienic treatments. Such men could not forget that the pathologic evidence often showed healing and that recoveries apparently did occur . . . Amidst all the contemporary pessimism, the phrase 'consumption curable' began to appear in some medical titles intended for a lay audience."

The changing death rates in the United States from tuberculosis, as compared with other infectious diseases, from 1900 to 1959, as well as the changing averages of life expectancy, are set

TABLE 3.
DEATH RATES FROM SEVERAL DISEASES AND LIFE EXPECTANCY IN THE CONTINENTAL UNITED STATES
Annual deaths per 100,000 of population

Year	Tuberculosis	Influenza, pneumonia[a]	Gastrointestinal inflammatory disease
1900	199	210	113
1950	15	24	4
1959	6	24	4

Average life expectancy at birth, years			
Year	Total population	Males	Females
1900	47.3	46.3	48.3
1950	68.2	65.6	71.1
1959	69.7	66.5	73.0

[a]Exclusive of the newborn.
SOURCE OF DATA: Lindsay and Allen (1961).

TABLE 4.
TUBERCULOSIS MORTALITY IN NEW YORK CITY, 1900-1961

	Males		Females	
Year	Deaths	Rate[a]	Deaths	Rate[a]
1900	5,783	338	3,847	222
1910	6,352	266	3,722	156
1920	4,211	149	2,924	103
1930	3,131	90	1,958	56
1940	2,394	65	1,233	33
1945[b]	2,448	65	1,065	27
1950	1,718	45	603	15
1952[c]	1,209	31	389	9
1955	840	22	244	6
1958	632	17	201	5
1961	564	15	174	4

[a]Per 100,000 population.　　　　[c]INH introduced.
[b]Streptomycin discovered.　　　SOURCE OF DATA: Lowell (1962).

out in table 3. The corresponding mortality rate from tuberculosis in New York City is given in table 4.

OCCULT "CURES"

The belief that disease is a form of divine punishment is found in much of ancient literature, including the Bible. As would be expected, attempts at the relief and cure of diseases often led to procedures of such a bizarre nature that they can only be described as *occult*.

The Roman naturalist Pliny (as quoted by Webb) noted these recommendations: "The cure for phthisis is effected by taking a wolf's liver boiled in thin wine; the bacon of a sow which has been fed upon herbs; or the flesh of a she-ass, eaten with the broth; the last mode in particular being the one that is employed by the people of Achaia. Smoke of dried cowdung, inhaled through a reed is remarkably good for phthisis." Webb adds: "Elephant's blood, and milk, especially a woman's and asses', were extolled as remedies. Affections of the lungs were supposed to be cured by eating mice boiled in salt and oil. An invaluable remedy for incipient phthisis was the root of consiligo."

Rolleston (1941), describing Christian practices in his paper on *The Folklore of Pulmonary Tuberculosis*, says: "Besides the Virgin Mary there are numerous patron saints for pulmonary tuberculosis whose aid may be invoked by sufferers from the disease," of whom the best known is "Pantellemon or Pantaleon who was physician to the Emperor Maximiamus. In the Middle Ages he was also the patron of physicians and midwives as well as one of the fourteen guardian martyrs. Other patron saints for the disease were Malo (Machutius, A. D. 565) and Bernulphus who was specially invoked for consumption in children. In Belgium a cure for consumption is to take every evening on going to bed a spoonful of blessed water with a pinch of the ashes of St. John's fire . . . Among charms connected with death or the grave may be mentioned that eating butter made from milk of cows fed in churchyards has been regarded as a sovereign remedy for consumption."

THE USE OF SALTS, INCLUDING THOSE OF GOLD AND COPPER

Louis in 1844 examined current methods of treating phthisis patients with iron salts, sodium chloride, potassium bicarbonate, ammonium hydroxide, calcium chloride, chlorine gas, digitalis, hydroorganic acid, creosote, and iodine, only to the conclusion that "the various means which have of late risen into notice, as possessed of the greatest power of effectually influencing the course of phthisis, or even of effecting its cure . . . have, one after another, vanished before scrutiny." He asserted, however, that this gave "no reason that we should despair for the future, or adopt the opinion that we shall never succeed in discovering some agent or other, capable of effectually opposing the onward course of phthisis once developed."

The use of gold preparations for the treatment of tuberculosis has been advocated at various times, particularly after Koch's finding that gold salts inhibit the growth of tubercle bacilli. Numerous preparations, under such names as "sanocrysin," "aurocantan," "krysolgan," were used. Copper was also recommended, and in Japan a compound of potassium cyanide and copper-cyanourate ("cyanocuprol") was used. These, like the gold salts, were found to be of little if any value in tuberculosis.

Among the above preparations, sanocrysin (sodium-gold-thiosulfate, 37.4 per cent containing gold) deserves special attention. Reports of its curative effects in tuberculosis were made by Møllgaard in 1924 on the basis of observations of the action of the compound on cultures of the organism, on experimental infections, and in the treatment of human cases by Danish clinicians. Upon introduction into the blood stream, the compound was said to permeate tuberculosis lesions and kill the tubercle bacteria. Amberson et al. in 1931 made a careful comparison of the effect of this preparation upon a group of patients, using another similar group as controls. They reported that on treating twelve patients with sanocrysin, no evidence could be detected of any beneficial effect upon pulmonary tuberculosis or its complications. As compared to the untreated "control" cases, those receiving the gold compound usually became worse, owing partly, at least in some cases, to the effect of the substance. The injurious effects were exerted on the gastrointestinal tract, the skin, the kidneys and other organs, and in the nutrition of patients.

Among the other compounds used in the treatment of tuberculosis, mention may be made of creosote, the virtue of which was attributed chiefly to its constituent cresols, and also guaiacol and its related compounds. Creosote, according to Long, was introduced as an antituberculosis agent about 1830. It has been called "the most used of the false specifics." Although many of these showed some bacteriostatic effects for tubercle bacteria *in vitro,* they had no therapeutic benefit in animals.

The Introduction of Sanatoria

One of the first modern weapons against tuberculosis, next to the pneumothorax, was the introduction of sanatoria. A botanical student from Silesia, Germany, Hermann Brehmer, himself a sufferer from the disease, was advised by his doctor to go to a better climate. Brehmer proceeded to the Himalaya Mountains, where he combined botanical research with a search for a cure for his disease. He returned home cured, and proceeded to study medicine. In 1854 he presented a thesis on the subject "Tuberculosis is curable." In the same year he built a house in Görbersdorf, in the middle of a pine forest, where those sick from tuberculosis could be well nourished and live in the open air. This was to become the model for all healing centers and sanatoria, a mighty force against pulmonary tuberculosis.

The further development of sanatoria had to await the final demonstration of the infectious nature of tuberculosis.

CONCLUSIONS

As the first half of this century was coming to an end, certainly at the beginning of the 1940's, no evidence whatsoever could be found to indicate that the chemotherapy of tuberculosis was possible. There was not even a ray on the horizon that would suggest such a possibility in the very near future.

Smith in 1941 succinctly summarized the results of the early and even some of the more recent efforts:

"Killing diseases have been fought with charms and herbs, with prayers and incantations, with royal proclamations and medical manifestoes, with nostrums and guinea-pigs. But tuberculosis is preeminently the disease which has been fought with publicity. It is 'tb' and 'the great white plague.' It has elbowed its way into the Christmas festivities. We 'stamp out tuberculosis.' The assault on it has set the pace for a dozen health crusades, and has held the spotlight longer than any of them."

The two important contributions made by all the early attempts to heal tuberculosis were (a) the search for better climate, which involved leaving the northern gloomy skies and departing for the south, and (b) proper rest, from which resulted the early efforts to provide suitable sanatoria. The idea of a sanatorium centered on treatment by fresh air and a healthful regimen had established the trend toward caring for tuberculosis patients in special institutions, and from this had come an important by-product—the carriers of the disease were segregated from the rest of the population.

Koch had predicted a fall in the tuberculosis death rate as a result of the isolation of the "polluting multitude" in sanatoria and hospitals. On the other hand, Smith believed that the gradual reduction in mortality from tuberculosis represented the waning of a long tuberculosis epidemic cycle. ". . . one can only speculate about it, but perhaps the decline in our own day should be thought of first of all as the downswing of a natural cycle, and only the next upswing will show how far we have managed to establish 'control.'" He suggested that our great-grandchildren be consulted for further details.

«Part II»

TUBERCULOSIS IN MODERN TIMES

«4»

The Diagnosis of Tuberculosis.
Laennec Discovers the Stethoscope.
Subsequent Discoveries

In the three thousand years up to the beginning of this century our understanding of the nature of tuberculosis, its proper diagnosis, and the methods of treatment has passed through five distinct periods:

1. The *ancient* or *classical* period of the Hebrews, Greeks, Romans, and other ancient peoples; the knowledge that has come down to us from that period is largely centered around or has been credited to Hippocrates, who often has been designated the father of medicine.

2. The *post-Renaissance* period, largely covering the sixteenth and seventeenth centuries; here the contributions of the great Italian, British, and other physicians (Fracastoro, Sylvius, Thomas Willis, Richard Morton, B. Marten) are outstanding.

3. The period of *clinical diagnosis,* covering the latter part of the eighteenth and the early part of the nineteenth century; the contributions to our knowledge of tuberculosis are centered here largely around the discovery of percussion by Auenbrugger and of the stethoscope by Laënnec.

4. The period covered by the latter part of the nineteenth cen-

tury, when the *infectious nature of tuberculosis was established;* this period encompasses the work of Villemin and Koch, and those immediately preceding or following them, the high point being the isolation of the causative agent of the disease by Koch.

5. Finally, the period of *accurate diagnosis and treatment,* covering the work of Röntgen on X-rays and Forlanini on the pneumothorax, and the development of sanatoria and bed rest.

Castiglioni (1933) has made a point of the fact that the history of medicine is not merely a catalogue of names of more or less celebrated physicians, or a "list of curious indications or the search for a grain of truth in the dusty volumes of ancient authors." Rather, he declares, we must understand that, "in its broadest sense, the history of medicine is a study of diseases: cause, diagnosis and treatment; the study of the marvelous progress from demonism to bacteriology; from the ancient organotherapy to modern serotherapy"—to which he might have added, "and, finally, chemotherapy."

EARLY AND LATER OBSERVATIONS

In the light of the above, one must recognize the contributions made at different historical periods to the solution of one of the most devastating diseases that have plagued the human race since time immemorial, and more particularly in modern times: tuberculosis. This can be done by noting the contributions of eminent clinicians and scientists (or philosophers in olden days). It is important to keep in mind, as Castiglioni emphasized, that the "observations of ancient writers should prove an efficacious means of undermining the thought current even among modern physicians, that tuberculosis is a relatively recent disease, and little known in antiquity . . . Cohnheim and Koch [have] stated that tuberculosis was a product of economic and social conditions, while it is far more correct to call it a disease of civilization which was observed in the remote past among the first groupings of people, progressing with the density of population in urban centers."

Hippocrates and his school were the first to describe tuberculosis of the lungs, under the name of "phthisis," although this concept also included empyema and "phyma," or abscess of the lung. The beginning of the disease was considered to take the form of a respiratory catarrh, a dry cough, accompanied by some yellow sputum, chest pains, increasing malaise, and later by fever. The characteristic weight loss was recognized, along with the other symptoms that became more pronounced as the disease progressed, such as evening fever followed by morning remissions, flushed cheeks, and dilated pupils. Long eyelashes are also mentioned, although modern physicians would probably doubt that such a cosmetic advantage had any relationship to tuberculosis.

In ancient Greece, as now, a sputum examination was considered essential in making a diagnosis. There were, however, basic differences. Hippocrates (Webb, 1936) looked for something other than tubercle bacilli: "If the sputum of a patient when poured on the coals has a heavy smell, the person is suffering from phthisis . . . when the expectoration of a consumptive emits a strong rancid smell in burning, and the hair falls off, the disease will be fatal. If expectorated matter sinks in seawater, the disease will be shortly fatal; but the water should be contained in a vessel of copper." The curability of phthisis in all stages was considered possible; it was believed that a change of residence was beneficial to the sufferer.

Not only was tuberculosis recognized by early physicians, but its contagious nature was also suspected. The famous philosopher of antiquity, Aristotle (384-322 B.C.) is credited by Webb with having sensed such danger. He regarded malaria as not being infectious, thus distinguishing it from plague and consumption. "Why, when one comes near consumptives, or people with ophthalmia, or the itch, does one contract their disease, while one does not contract dropsy, apoplexy, fever, or many other ills? . . . [With] the consumptive one breathes this pernicious air. One takes the disease because there is in this air something disease-producing."

Aretaeus (A.D. 50) gave a clear description of tuberculosis, which he differentiated from empyema; sea voyages and country air were believed to be beneficial. Pliny the Elder recommended pine forests for their healing powers. The Romans were aware not only of the contagious nature of tuberculosis and that some patients recovered, but also of the fact that it is a relapsing disease.

A century or so later, Galen stated that it is dangerous to live with consumptives. He advised patients to go to the highlands. Anhyllus, in the fourth century, also recommended a "cure of altitude" for consumptives, and approved of sun bathing and sea voyages.

Subsequent to these classical times, no further progress was made in the clinical understanding and diagnosis of tuberculosis, or in treatment, until the sixteenth and seventeenth centuries. Then, according to Shryock (1957), "Some definite knowledge was acquired of the pathology of both phthisis and scrofula, and their relationship began to be suspected. The possibility of some connection between pathology and symptoms was also beginning to receive attention. In the case of phthisis, finally, better understanding was achieved of its complex, clinical behavior. The major area in which no progress seemed to be made was that of etiology, and the lack of knowledge in this field prevented both the identification of the total disease and any agreement on the means of its transmission."

Sydenham suspected the possibility of infectious agents, says Webb, for he wrote: "Certain diseases are caused by particles which are disseminated throughout the atmosphere [and] which become mingled with the blood and thus are distributed throughout the entire organism." Morton thought that the tubercles were the true cause of phthisis; he believed in the hereditary nature of the disease, as well as its contagious properties. It was Sylvius (1478-1555), however, who was the first to emphasize clearly the connection between tuberculous nodules and phthisis. These nodules were regarded as enlarged lymph nodes in the lung, analogous to scrofula. The inheritance of phthisis depended on the scrofulous constitution. He described the symptoms of the disease and believed in its contagious nature. Morgagni (1682-1771) questioned the relationship of the tubercles and nodes, and believed that phthisis could originate from other causes; he regarded the disease as highly infectious, and went so far as to refrain from doing autopsies on consumptives.

Tuberculous meningitis was first recognized in the eighteenth century by Robert Whytt (1714-1776), an Edinburgh neurologist. The symptoms of tuberculosis were described by John Fothergill (1712-1780), who observed that the disease attacked particularly the young, although it could also occur in adults. At autopsies a "clear pellucid lymph" was found in the ventricles. According to Webb, one in six of the deaths in London in Fothergill's time was due to "phthisis pulmonalis," which led foreigners to claim that consumption was "the grand epidemic of England."

DISCOVERY OF PERCUSSION

In 1761, a Latin pamphlet of less than fifty pages was published, under the title *Inventum novum* . . . , or, in English, *"A New Discovery of Percussion of the Human Chest for Detecting the Signs of Obscure Disease of the Chest Cavity."* The author, Auenbrugger, said: "I present you, dear reader, a new method of discovering diseases of the chest invented by me. It consists of knocking at the human thorax. From the variation in reverberating sounds a judgment may be reached as to the condition of the contents of the chest cavity. Neither a craving for authorship or an outburst of speculation, but seven years of observation is my reason for classifying, putting in order and publishing what I have thus discovered." (Quoted in Flick, 1925.)

This discovery, or as some consider it, rediscovery, of the diagnostic value of percussion, is believed to be the first advance in the diagnosis of pulmonary disease since the days of the Greco-Roman physicians. It was highly esteemed by Laënnec, who stated in the introduction to his own book (as quoted in

Flick): "I feel confident that physicians who have regularly and for some time been engaged in opening dead bodies will agree with me when I say that before Auenbrugger's discovery one-half of the peripneumonias and acute pleurisies and almost all the chronic pleurisies must necessarily have been overlooked . . . Percussion of the chest according to the method of the ingenious observer whom I have just cited is unquestionably one of the most valuable discoveries with which medicine has ever been enriched."

Laënnec (fig. 6) knew that the ancient physicians made use of touch, inspection, measurements, succussion, and even immediate auscultation. He himself did not claim priority in looking beyond subjective symptoms for physical evidence of disease in diagnosis.

During the first part of the nineteenth century definite progress was made in the early diagnosis of tuberculosis. In his *A Practical and Historical Treatise on Consumptive Diseases* in 1815 Thomas Young mentioned the early symptoms of general malaise, slight shortness of breath, and cough without pain; only one in a hundred was said to recover from the disease.

Bayle, who died of consumption in 1823, at the age of forty-two, probably contracted his infection at the autopsy table. It may be noted here that many pathologists of that period died of tuberculosis. Bayle reported, in 1810, that there were six kinds of pulmonary consumption, of which chronic ulcerative tuberculosis was by far the most common. The other five pulmonary lesions (and their modern equivalents) were "granular" (miliary tuberculosis), "melanotic" (anthracosis), "ulcerous" (pulmonary abscess and gangrene), "calculous" (encapsulated, calcified tubercles) and "cancerous" (true tumors). Bayle drew attention to the relation of chronic ulcerative tuberculosis of the lungs to tuberculosis of other organs, such as the larynx, intestine, and mesenteric lymph nodes.

Laënnec, however, in his *De l'auscultation médiate* of 1819, held that there was only one form of phthisis, namely, that in which the tubercle was an "accidental production." He observed the successive eruptions of tubercles which might be found in the same lung. The tubercles could also occur in various parts of the body, and no part of the body was immune. Laënnec traced the disease through all its pathological forms from the

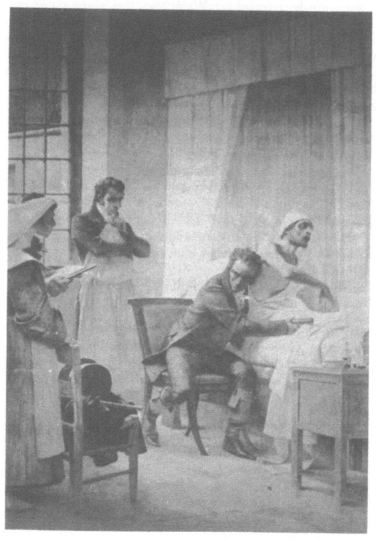

FIG. 6. René Théophile Hyacinthe Laënnec, examining a patient.

tiny tubercle to the larger necrotic cheesy tubercle and in all the organs of the body commonly attacked by tuberculosis.

Although Laënnec did not accept the idea of contagion in tuberculosis, he reported, as Webb says, that phthisis was contagious "in the opinion of a few laymen and a few physicians of the south." He infected himself accidentally, about 1800, when sawing a tuberculous vertebra during an autopsy. Webb considered this to be "the first clearly described account of transferred tubercle infection." Laënnec also believed that tuberculosis, even in the cavity stage, was curable.

Castiglioni notes that in his "search of the incidental cause of tuberculosis," Laënnec maintained his opinion that changes of temperature and atmospheric conditions were in general, "merely the causes for a more rapid development of tuberculosis, assuming, however, the preëxistent condition of the malady . . . " Also, Laënnec pointed out that tuberculosis was most frequent in urban centers. He recognized that "hemoptysis is never the cause of phthisis, but simply the consequence of a previously existent pulmonary affection." He also observed that mental depression, particularly if of long duration, was an important factor in predisposition to tuberculosis. Regarding heredity, he held that predisposition was hereditary, but not the disease itself.

Discovery of the Stethoscope

Laënnec outlined his work dealing with the invention of the stethoscope as follows (quoted in Rapport and Wright, 1952): "In 1816 I was consulted by a young woman presenting general symptoms of disease of the heart. Owing to her stoutness little information could be gathered by application of the hand and percussion. The patient's age and sex did not permit me to resort to the kind of examination I have just described (i.e., direct application of the ear to the chest). I recalled a well-known acoustic phenomenon, namely, if you place your ear against one end of a wooden beam the scratch of a pin at the other extremity is most distinctly audible. It occurred to me that this physical property might serve as a useful purpose in the case with which I was then dealing. Taking a sheaf of paper, I rolled it into a very tight roll, one end of which I placed over the praecordial region, while I put my ear to the other. I was both surprised and gratified at being able to hear the beating of the heart with much greater clearness and distinctness than I had ever done before by direct application of my ear."

Laënnec dealt with the treatment of tuberculosis in greater detail. He stated (as quoted in Flick): "To conceive of the possibility of curing it in some cases after the formation of an ulcerous cavity of the lung is a matter which seems to be simple enough to most practising physicians who are not anatomists, but which nevertheless appears to be absurd to most of those who have given themselves over with persistence to research in pathological anatomy . . ." He saw the evidence of this in the innumerable remedies proposed for the disease. In treatment, it was important to recognize the disease early and to prevent secondary eruptions of tubercles. He had little confidence in bleeding, since it could neither prevent the development of tubercles, nor cure them when they were formed, and held that it "should not be used in the treatment of pulmonary phthisis except to allay an inflammatory complication or an acute sanguinary congestion." He thought favorably of change of climate and of the use of milk.

Laënnec stated further: "Before the character and progress of the development of tubercles was well known and when phthisis generally was attributed to a chronic inflammation and to a mild suppuration in the lung tissue, physicians no more than the public, doubted the possibility of curing it by proper treatment, especially when this was begun in time and before the disease had reached its first stage. Now, nearly all men in the profession who have kept in touch with recent progress in pathological anatomy, think that the tuberculous disease is like the cancerous affections, absolutely incurable." He was convinced, however, that "in a great many cases of pulmonary phthisis one can still hold out the hope, as in the examples which we have reported, of real recovery or at least of a suspension of the symptoms and burdens of it which is nearly the same thing, since the patient can be restored to a condition of health sufficiently perfect to enable him to perform the functions of civil life for many years before the development of the tubercles, still in a state of crudity, brings on a new and last attack of phthisis."

Flick evaluated the contribution of Laënnec to our knowledge of tuberculosis as follows: "By his discovery of auscultation he not only gave to the clinician a new resource for diagnosis but he made its twin percussion, more valuable. Up to his time percussion had not been used much, although his colleague, Dr. Bayle, used it regularly and his teacher, Corvisart, taught it and promulgated it by translating Auenbrugger's work into French. He likewise made percussion and auscultation a valuable handmaid to pathology by testing out, as Auenbrugger had done, the correctness of his clinical findings with the exact conditions which the autopsy revealed."

According to Shryock, "Although Laënnec, like all scientists, followed earlier leads, there is some justification for the view that his work marks a turning point in the history of tuberculosis. He clearly identified the disease entity, admitted the futility of known therapy, and emphasized the only available means of palliative treatment. His unitary concept facilitated later investigations in etiology. Yet in some measure, this recognition of Laënnec's achievements is a judgment after the fact. At the time, the French clinician's position was by no means generally accepted . . ."

Laënnec dedicated his life to the study of the diagnosis and nature of tuberculosis. Unfortunately, this great scientist and clinician, who opposed the theory of the contagious nature of the disease, contracted it himself in his early youth. Returning to his beloved Brittany, he died there of the disease in 1826. He described his own condition, according to Kervran (1960), as follows:

"Strength diminished by half since Paris, noticeable loss of weight *idem*. More or less perceptible feelings of fever, particularly during the day. Practically no more perspiring. Expectoration more abundant rather than less, always the same: The fistula of the maxiliary sinus gives of a 'gros' (about 0.5 gm) of pus every twenty-four hours; obviously there is decay of the perforated alveolus. The large intestine appears to be healthy and no pain is felt from emollient enemas of one and a half litres. Appetite is more or less the same . . ."

It seems that to his dying days he refused to renounce life. He was still considering buying some more land. This, as Kervran says, was possibly an effect of the final euphoria of tuberculosis.

Laënnec's ideas were not universally accepted. No less a person than the "Pope of Medicine," Robert Virchow, wrote (as quoted in Flick): "The great and well-founded authority of Laënnec in this matter has prejudiced the entire coming generation, and has tied up the whole development of the teaching about tuberculosis in a one-sided way to one organ, which . . . is really the most difficult and the least available for the study of these processes. All the conditions of the lungs, which are accompanied by the formation of cheesy masses in their tissues, as well as those with ulceration of the parenchyma, have been summarized under the conception of tubercle, and in this way may have been given an unnatural unity of which the conflicting elements have manifested themselves most clearly in the manifold differences about the nature and composition of the changes." Until his death, even after Koch's classical work on the tuberculosis organism, Virchow still spoke of "the so-called tubercle bacilli."

Judgment has been well given by Krause (1932), who has said that Laënnec's *Traité de l'auscultation médiate*, Villemin's *Etudes sur la tuberculose*, and Koch's *Die Aetiologie der Tuberkulose* constitute "the tripod supporting modern medical method as this scours the great field of infectious diseases."

MICROSCOPIC AND CULTURAL EXAMINATIONS

For the diagnosis of tuberculosis, the stethoscope was later supplemented by the microscopic and cultural examination of the sputum for the tuberculosis organism. In staining tubercle bacteria the Ziehl-Neelsen technique has been most commonly used. The process was introduced by Paul Ehrlich, who stained the sputum with a mixture of gentian violet or fuchsin and aniline, and then treated it with nitric acid. This type of stain became known as "acid-fast," since it was not discolored by the nitric acid. It was developed further by Ziehl, who substituted carbolic acid for aniline (later using carbol-fuchsin), and by Neelsen, who used fuchsin and sulfuric acid instead of gentian violet and nitric acid. By this method the acid-fast bacteria are stained red and the other organisms, as well as the tissue cells, blue.

Rontgen Discovers the X-rays

Another fundamental contribution to the means at our disposal for the diagnosis of pulmonary disease was made by Röntgen in 1895 with his discovery of the X-rays. Later refinements, such as the fluoroscope, and stereoscopic machines, made this discovery one of the major advances in modern-day diagnosis of tuberculosis. Although this discovery was made subsequent to Koch's work, it may be mentioned here, since it is the final step in the early and accurate diagnosis of tuberculosis.

Brown (1941) divides the evolution of the X-ray method of exploration of tuberculosis into four periods: the discovery and pioneer period (1895-1905); the formative period (1906-1915); the period of expansion (1916-1925); and the period of realization (1926-1935). He says:

"Late in October of the year 1895, Wilhelm Conrad Röntgen, Director of the Physical Institute of the University of Würzburg, decided to make certain experiments with the cathode rays. He set up his apparatus, consisting of a Ruhmkorff induction coil with a mercury interrupter and [a] vacuum tube. The room was darkened and, as he was attempting to pass a high-tension current through the tube to test the density of its black cover, he suddenly saw a few brightly fluorescent crystals which lay upon a table at some distance from the tube. This observation, which later proved to be so significant, was made on a Friday night 'at a late hour when assistants were no longer to be found.' The discovery of these rays, which were capable of passing through opaque objects, opened the door for the examination of bodies to determine their physical structure. Here was a ray which would reveal in shadows that which was hidden from the eye . . . [The first] radiographs were made of the human body during the early part of January, 1896 . . . "

FORLANINI AND THE PNEUMOTHORAX

In 1882, the same year that Koch published his epoch-making contribution on the tubercle bacillus, Forlanini published a paper entitled "Contribution to the Surgical Therapy of the Ablation Phthisis? Ablation of the Lung? Artificial Pneumothorax?" Because of the great influence of Koch's ideas in the field of tuberculosis, there was at first indifference and even considerable misunderstanding of the work of Forlanini. It was not until thirty years later that the artificial pneumothorax was given official recognition and official blessing of the scientific world, at the International Congress of Tuberculosis held in Rome in 1912. Today, when chemotherapy of tuberculosis has attained a high pinnacle, the artificial pneumothorax has lost its place in the treatment of pulmonary tuberculosis, with the possible exception of those cases where the patient is resistant to chemotherapy.

Forlanini began the intrapleural injections of nitrogen in a patient with hydropneumothorax. The results were encouraging. This work was briefly reported in the *Medizinische Wochenschrift* of Munich in 1894 (as given in Brown):

"Based on the repeated favorable influence of pleural fluid and of pneumothorax in the course of phthisis, Forlanini had made a research on the living and treated two cases with favorable results. The pneumothorax can naturally be used only on one side. The pleura bears the presence of air well [and no] reaction or irritation or any harmful influence [was] observed. Oxygen was more quickly absorbed than nitrogen which is now used to produce pneumothorax. The operation must be slowly and gradually performed [and] each day a small quantity of nitrogen introduced, [of which,] if it be absorbed, the quantity must be increased. The absorption capacity of the pleura lessens as time goes on. After some months the interval between injections can be lengthened."

True surgery as well (thoracoplasty and resection) continued to make contributions toward the alleviation of tuberculosis. Concerning Forlanini's and later work, Brown wrote in 1941: "There are few achievements of surgery that can surpass what has been accomplished in the last thirty years in the surgical treatment of tuberculosis . . . The foundations have been well and truly laid. Most of the superstructure had already been added. What remains to be done cannot be much more than detail work."

Thus, at the end of the nineteenth century, knowledge of the cause of tuberculosis, occurrence, diagnosis, general care and certain important methods of surgical treatment had become generally recognized. Another half century, however, was to elapse before the coming of chemotherapy, which, if properly used, should deliver the last blow in the control of the disease.

« 5 »

Villemin and Koch Establish the Role of the Tuberculosis Organism as the Causative Agent. Further Attempts at Prevention and Treatment

The isolation of the causative agent of tuberculosis had to wait until the latter part of the nineteenth century. J. A. Villemin (fig. 7) first demonstrated in 1865 that phthisis is a specific affliction; that the cause of it was an inoculable agent or germ; that it was transferable from man or cow (bovine tuberculosis being more virulent than human) to the rabbit and the guinea pig; that it could also be transmitted from one infected animal to another in continued series; and that tuberculosis belonged among the virulent diseases and should be classified with smallpox, scarlet fever, syphilis, or, better still, with glanders.

Villemin's first important communication was his paper in 1865, "On the Cause and Nature of Tuberculosis and Its Inoculation from Man to Rabbit." As reported by Castiglioni (1933), he demonstrated that "tubercularization of animals may be obtained with the injection of sputum or blood from tuberculous rabbits, using control experiments. On the other hand, he had negative results in inoculation with carcinomatous—fibrous or

FIG. 7. Jean-Antoine Villemin.

soft—substance, and from these experiments he definitely formulated the thesis that the cause of tuberculosis must be looked for
exclusively in a specific virus and that this virus is the only
cause of tuberculosis. Contrary to all that [had] been previously
said, Villemin firmly maintained that tuberculosis does not have
its spontaneous origin in the organism and that neither emaciation, misery, atmospheric disturbances, heredity, occupations nor
progressive maladies stand in direct causal relation to tuberculosis. The cause of tuberculosis is a germ, the bearer of specific

tuberculous virus, which is to be found in the air. The possibility of transmission of this germ is shown by the fact that it lives and transplants itself in organic tissue of man and of certain animals."

In his book *Études sur la Tuberculose,* in 1868, Villemin noted that "tuberculosis was more frequent among the medical personnel and soldiers stationed for long times in barracks than among troops in the field," that "healthy young men from country districts often became consumptive within a year or two after their arrival in army posts," and that "prisoners, industrial workers and members of religious cloistered orders were more apt to contract the disease than were ordinary civilians."

Villemin noted further that tuberculosis flourishes only rarely in high altitudes and that it increases in crowds and occurs most frequently in large manufacturing and commercial cities; that it afflicts persons who live in common and who are confined indoors, such as prisoners, members of religious orders, and soldiers; that it spares those who are not in close contact with others or who live in the open air in a nomadic or savage state; that although it is common among troops in barracks, it ceases to be so when the soldier is on campaign and is not housed; that the dwelling together of persons in close, badly ventilated houses is followed by tuberculization of many of them; that phthisis, unknown among certain peoples of America and Oceania, became the most violent, destructive scourge among them when they came in contact with Europeans; that bovine phthisis, like human, increases with confinement and crowding of the animals, affecting a large number of them when it breaks out in a stable; that recognition of the contagiousness of phthisis has been maintained at all times by popular belief and by observers of the greatest merit; that tuberculosis develops and propagates itself according to an ensemble of conditions, analogous to those of zymotic diseases. The conclusion was thus reached that the cause of this disease resides in a specific agent, which multiplies and transmits itself under the above circumstances.

Villemin's conclusions "excited widespread discussion and control experimentation, for such far-reaching statements commanded attention," according to Baldwin (1913), who adds: "A new era of science was founded about the same time by Pasteur, and had he not become engaged in other investigations it is permissible to think that his genius might have crowned the work of Villemin by anticipating Koch's discovery by several years."

Villemin demonstrated that there is a difference in the behavior of human and bovine tuberculosis: "We must remark that none of our rabbits inoculated with the human tubercle have presented a tuberculization so rapid and generalized as that which we have obtained with the tubercle inoculation of the cow."

Villemin thus proved conclusively the inoculability of tuberculosis from man to animals and from animal to animal. This was true for material obtained both from lymph nodes and from lung tuberculosis, leading to the conclusion that the causative agent was the same. He further concluded that this agent was a germ which bore a specific tuberculosis virus; it was believed to come originally from the surrounding air.

Villemin proposed a new concept of the relation between tuberculosis and scrofula. Characteristic manifestations of tuberculosis were obtained in cases where he had inoculated matter from a scrofulous gland; he suggested that all forms until then designated scrofula belonged to the general type of tuberculosis.

For combating tuberculosis, Villemin recommended the improvement of habitations and working conditions, the maintenance of a high standard of health, and, if possible, a return to the practice of disinfection of the things and places which may possibly have been contaminated by consumptives. He stated, according to Flick (1925): "Now that we know the nature and the intimate cause of tuberculosis, that we can produce it in animals at our pleasure, does not the horizon rise up before us full of consoling hope? Do not numerous and repeated experiments, which one cannot make on man, promise some results? We modify smallpox by vaccinia and give an artificial immunity against the virus of smallpox; is it forbidden us to hope that the tuberculous virus will some day reveal its antagonistic substance? And how immense would be the results from a neutralizing agent, applied in the beginning of the disease, and destroying in the organism the morbid principle which there multiplies itself! For the danger in phthisis is not in a few tubercles, which often disturb the respiratory functions so little that the patient does not experience any effect and the most careful auscultation scarcely recognizes them, but in the impregnation of the entire economy with the disease-producing substance, bringing about successively tuberculous extensions more or less close together and abundant, which lead unfortunately to death. These are the ideas which should direct us in the research for curative and prophylactic measures in tuberculosis. These are the hopes which our discovery luringly had held out for us. May they bring the fruit to which we look forward in the future with pleasure!"

Practically simultaneously with Villemin, Budd recognized in 1867 that tuberculosis is a specific disease, in the same sense as typhoid fever, scarlet fever, typhus, and syphilis are. Tuberculosis never originates spontaneously; it is perpetuated solely by the law of continuous succession. The tuberculous matter itself constitutes the material by which phthisis is propagated from one person to another and disseminated through society. Budd suggested that, by the destruction of this matter on its issue from the body, by means of proper chemicals, combined with good sanitary conditions, we might hope that eventually, and possibly at no very distant future, to rid ourselves entirely of this scourge.

Rivolta and Perroncito in a report of their work *on the structure and contagion of tuberculosis in animals,* in 1868, described lesions they had produced in rabbits and guinea pigs by grafting tuberculous substance from cattle. They demonstrated that in animals tubercles and tuberculous nodules are identical in structure and that they undergo the same process as those which characterize tuberculosis in cattle and man, thus confirming their histological identity.

Klebs, in 1877, upheld the specificity of the tuberculosis "virus" and its identity with the bovine form, the sputum being looked upon as the "virus" carrier. Klebs was the first actually to transfer the organism by artificial culture on egg albumen through several

FIG. 8. Robert Koch.

generations before inoculation. He did not, however, recognize the true nature of the tubercle bacillus. He found a motile organism, designated as *Monas tuberculosum,* which he considered to be the *contagium vivum.* Although others, independent of Koch, described bacilli to be present in the center of the tubercles, they did not cultivate the organism, nor did they stain it; they could hardly claim, therefore, to have identified the active agent responsible for the disease. This remained to be done by Koch.

KOCH ISOLATES THE TUBERCLE BACILLUS

The climax came in 1882 when Robert Koch (fig. 8) presented irrefutable proof that a specific microbe was the basic cause of tuberculosis. In a dramatic presentation before the Physiological Society of Berlin, on March 24, 1881, of his paper on *The Etiology of Tuberculosis,* he said in part:

"Villemin's discovery that tuberculosis is transmissible to animals has, as is well known, found varied confirmation, but also apparently well-grounded opposition, so that it remained undecided until a few years ago whether tuberculosis is or is not an infectious disease. Since then, however, inoculations into the anterior ocular chamber, first performed by Cohnheim and Salomonsen, and later by Baumgarten, and furthermore the inhalation experiments done by Tappeiner and others have established the transmissibility of tuberculosis beyond any doubt, and in future tuberculosis must be classed as an infectious disease . . . The aim of the study had to be directed first toward the demonstration of some kind of parasitic forms, which are foreign to the

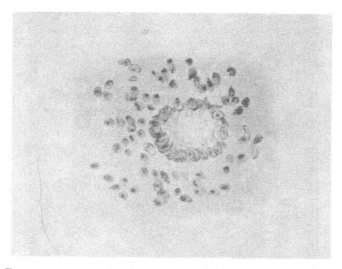

FIG. 9. One of Koch's tissue sections from caseated bronchial lymph node of a case of miliary tuberculosis; radially disposed bacilli in giant cell. Koch's colored plates show the bacilli as brilliant blue rods set in a background of all other elements stained by vesuvin.

body and which might possibly be interpreted as the cause of the disease. This demonstration became successful, indeed, by means of a certain staining process, which disclosed characteristic and heretofore unknown bacteria in all tuberculous organs." The

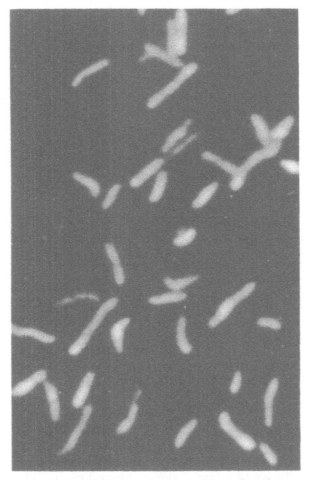

FIG. 10. *Mycobacterium tuberculosis,* under fluorescense microscopy and stained with auramine and rhodamine. (AFTER TRUANT, BRETT AND THOMAS, 1962.)

causative microscopic agent came to be known as *Mycobacterium tuberculosis* (figs. 9, 10, 11), although, in French-speaking countries, it is still spoken of as Koch's bacillus. In common parlance it is called the "tubercle bacillus."

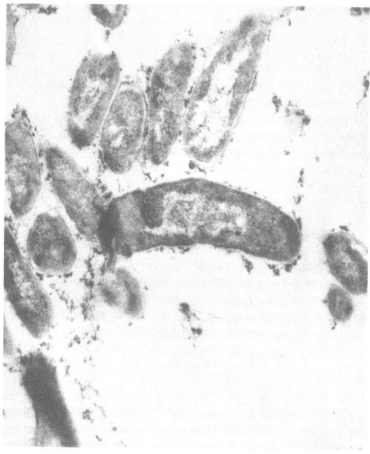

FIG. 11. Electron micrograph of a thin section of *Mycobacterium tuberculosis* ($\times$ 90,000). (COURTESY OF FLOYD M. FELDMANN AND GEORGE CHAPMAN.)

Koch thus definitely established that tuberculosis was to be attributed not to "bad blood" or "bad heredity," but to a specific organism, the *M. tuberculosis*. Koch at first failed to grow the organism on artificial culture media. He then used the method of Tyndall of sterilizing ox and sheep sera at 58° C. and coagulating at 65° C. When the media were later examined with a magnifying glass, Koch noted a growth of the tubercle organism before the end of the first week. When the culture was inoculated into laboratory animals, tuberculosis was reproduced. Koch had thus established that a specific bacterium was the cause of tuberculosis. This organism could be cultivated from the infected tissues. When inoculated into an animal, it caused the specific disease and could again be recovered.

The audience that listened to Koch's paper greeted its conclusion with "profound and utter silence," according to Rapport and Wright (1952). "His opponent Virchow—one of the leaders in modern medicine, the founder of cellular pathology—who had attacked the idea of the specificity of tuberculosis, left the room without attempting a reply. Paul Ehrlich, who was to discover salvarsan and who was also present, wrote later, 'That evening remains graven in my memory as the most majectic scientific event in which I have ever participated.'"

According to Dubos and Dubos (1952), "Villemin suffered much in his pride from seeing his work contemptuously ignored by Koch and all but forgotten by the rest of the world. He would have been wise to accept the cruel law of scientific life: 'He becomes the true discoverer who establishes the truth: and the sign of the truth is the general acceptance . . . In science the credit goes to the man who convinces the world, not to the man to whom the idea first occurs.'"

The bitterness felt by Villemin is illustrated in a letter he wrote to Pasteur in 1887, quoted by Cummins (1949): "I do not hope to gain a place beside you but—you will see that I am less modest than I appear—I have been so much discussed, so often attacked, that I suffer a certain amount of distress in thinking that the leading scientific academy still gives, at least, a sort of toleration to my former enemies. You must well remember the discussion and controversies of 1867 and 1868 both in the learned societies and in the press of France and other countries. Twenty years have passed and all this has become merely ancient history! Koch's bacillus, of which the Germans are so proud, has obscured the memory of what the French scientists had already accomplished. Koch will enter the Académie des Sciences through widely flung doors, in the triumphant way that has made a conquest for him of all the honours of his country."

Koch indicated in his paper the direction that preventive measures should take: "Tuberculosis has so far been habitually considered to be a manifestation of social misery, and it has been hoped that an improvement in the latter would reduce the disease. Measures specifically directed against tuberculosis are not known to preventive medicine. But in future the fight against this terrible plague of mankind will deal no longer with an undetermined something, but with a tangible parasite, whose living conditions are for the most part known and can be investigated further. The fact that this parasite finds the conditions for its existence only in the animal body and not, as with anthrax bacilli, also outside of it under usual, natural conditions, warrants a particularly favorable outlook for success in the fight against tuberculosis. First of all, the sources from which the infectious material flows must be closed as far as this is humanly possible. One of these sources, and certainly the most essential one, is the sputum of consumptives, whose disposal and change into a harmless condition has thus far not been accomplished."

No attempt will be made here to describe the organism causing tuberculosis or to emphasize the relationship between the human, bovine, and avian forms, or the relation of the mycobacteria to closely related bacteria. One aspect, however, must be mentioned here, since it has an important bearing upon the present discussion, namely, the epidemiology of the disease.

The tremendous effect of Koch's work upon both the lay public and the medical world has been described by Dubos and Dubos: "It was immediately regarded as a landmark, indeed as heralding a new era in the study and control of the disease. Yet it did not propound a new conceptual scheme, for the germ theory of disease was already established. It did not offer a new approach to the problem of tuberculosis, for by 1882 the discovery of the microbial agent of the disease was considered only a question of time. The isolation of the tubercle bacillus did not require a new experimental methodology, for Koch merely had to modify details of techniques already worked out largely as the results of his own earlier work. Men were stunned by the practical import of the event, rather than by the intellectual effort involved in its genesis. What electrified the world was not the scientific splendor of the achievement, but rather the feeling that man had finally come to grips with the greatest killer of the human race. 'The captain of all the men of death' was no longer a vague phantom. The heretofore unseen killer was now visible as a living object, and its assailants at last had a target for their blows. In Europe and America Koch became the pope of medical science. In Japan a new shrine was dedicated to him, as to a demigod."

When the implication of Koch's discovery was universally recognized, there was a general feeling that since tuberculosis had been shown definitely to be a contagious disease there was no reason why it could not be prevented. As time went on, however, it became evident that the problem was not so simple.

What followed has been told succinctly by Castiglioni: "The dream of all who witnessed Koch's discovery—to find a specific remedy in the treatment of tuberculosis—has not been realized, and the history of therapeutic experiments of the [subsequent] four decades is a history of great hopes and profound disillusions. Errors and misconceptions followed in the wake of Koch's discovery, and dominated the medical world for several years. Scientific workers ceased to regard tuberculosis as a clinical factor or . . . as a series of anatomico-pathological and clinical facts worthy of observation, and considered it almost exclusively from the viewpoint of bacteriology. It was an epoch in which medical men in the university clinics believed they could learn [and] clinicians believed they could teach the nature and the treatment of tuberculosis within the four walls of the laboratory; [in which] the increasingly diligent use of the microscope and dyes was more important than visiting hospital wards, and placing an ear to the patient's thorax."

Koch marred his own record considerably. In 1890 he announced that he had discovered a substance that could be used as a protective against tuberculosis; it was even said to be a remedy for existing tuberculosis. This new "discovery" was presented before the Tenth International Congress of Medicine in Berlin. It caused universal rejoicing. Koch refused at first to reveal the nature of the particular substance, which was referred to as "Koch lymph." He later disclosed that it was a glycerine extract of the tubercle bacillus, the product now known as "tuberculin." This announcement was greeted in all the leading medical publications. The English journal *Lancet* welcomed it in an editorial as "glad tidings of great joy." Soon, however, Koch was being condemned "for his unethical behavior in having attempted to keep secret the composition of the substance, and for his lack of scientific judgment in having recommended it as a remedy . . ."

Dubos and Dubos have explained this phenomenon as follows: "The reasons which made Koch try to keep secret the nature of his 'remedy' have never been made public. On several occasions Koch stated that the preparation of his material and the technique of its use required such skill that therapeutic accidents were likely to occur if the method were made available too soon to untrained practitioners . . . It seems, however, that the policy of secrecy was dictated by the German Ministry of Health. There were rumors, according to *Lancet* of 1890, pointing to the intention of the German Government to retain, as it were, the monopoly of the remedy in its own hands . . . to become the proprietary of a vast patent-medicine factory."

Others have found both Koch's premature enthusiasm and his secrecy understandable in view of the slow progress that had been made in treating tuberculosis, as compared with other infectious diseases. The case has been well put by Shryock (1957): "The logic was good but the result unsatisfactory . . . [Koch's] authority was such as to provide wide publicity, and much hope was aroused all over the world. But ensuing experience showed that the employment of the vaccine . . . was useless or even dangerous. Eventually, it came to serve a diagnostic purpose in the 'tuberculin test,' and further extensive experiments in vaccination were made in time. Yet the outlook in this area was not hopeful in 1900."

The Development of Sanatoria

One of the eminent followers of Koch in the United States was Edward L. Trudeau, who was stricken with tuberculosis in 1872 while nursing his brother. He consulted a physician because of the swelling of the lymph nodes along the side of his neck. No one knew at that time that scrofula is tuberculosis of the lymph nodes of the neck. Trudeau's general debility led his physician to advise the use of plenty of bacon for breakfast and to prescribe an iron tonic and painting of the lymph nodes with iodine. He had suffered occasional attacks of fever, which he attributed to malaria, a very common disease at that time. He finally consented to be examined, and learned that he had tuberculosis.

At the College of Physicians and Surgeons, where he took his medical training, he was taught, as he said in his autobiography (1916), that tuberculosis "was a non-contagious, generally incurable and inherited disease, due to inherited constitutional peculiarities, perverted humours and various types of inflammation," and where instruction "dwelt at length on the different pathological characteristics of tubercle, scrofula, caseation and pulmonary phthisis." From his previous experiences with his older brother, Trudeau feared that a similar fate, "speedy death," was awaiting him.

When Trudeau learned that this disease was not necessarily

fatal, provided he submitted to proper rest, he retired to the Adirondack Mountains "in order to lead an open-air life in the great forest, alone with nature." His process of reasoning was somewhat as follows: "If I had but a short time to live I yearned for surroundings that appealed to me and it seemed to meet a longing I had for rest and the peace of the great wilderness."

Impressed by "the difficulty of obtaining suitable accommodations in the Adirondacks for patients of moderate means," he established in 1885 the Adirondack cottage sanatorium,[1] that later came to be known as the Trudeau Sanatorium. Twenty-five years later, more than 400 such sanatoria were to be found throughout the United States.

Trudeau also undertook a comprehensive study of the tuberculosis culture and possible methods of preventing the disease. In referring to Koch's paper, he said: ". . . every step [was] proved over and over again before the next step was taken, and the ingenuity of the new methods of staining, separating and growing the germs read like a fairy-tale to me . . . I decided the next time I went to New York to devote all my efforts to learning to stain and recognize the tubercle bacillus."

This was the beginning of the Saranac Laboratory, based on the idea of producing immunity against tuberculosis by using killed cells of the tuberculosis organism. Said Trudeau: "I began to realize about this time that the direct destruction of the germ in the tissues by germicides was a hopeless proposition and, inspired by Pasteur's work on anthrax, chicken cholera and hydrophobia, I sought to produce immunity in my animals by dead germs, *or preventive inoculations of substances derived from the liquid cultures from which the bacilli had been filtered*. I published this work . . . as early as November 22, 1890, describing

---

[1]From the Latin word *sanare*, meaning to cure or heal.

Among the first hospitals for consumption was the Brompton Hospital in London, founded in 1841. Special wards were later established at Bellevue Hospital and the City Hospital (Blackwell's Island) in New York City. The Chestnut Hill Hospital for Consumptives was opened in Philadelphia in 1886.

my experiments in detail and giving my conclusions that neither the dead germs nor the soluble poisonous substances derived from liquid cultures of the tubercle bacillus protected rabbits and guinea pigs against subsequent inoculations."

In 1908, in a greeting to the delegates to the International Congress on Tuberculosis, Trudeau said: "For thirty-five years I have lived in the midst of a perpetual epidemic, struggling with tuberculosis, both within and without the walls, and no one can appreciate better than I do the great meaning of such a meeting. I have lived through many of the long and dark years of ignorance, hopelessness, and apathy, when tuberculosis levied its pitiless toll on human life unheeded and unhindered; when, as Jaccoud has tersely put it: 'The treatment of tuberculosis was but a meditation on death.' But I have lived also to see the dawn of the new knowledge, to see the fall of the death-rate of tuberculosis, to see hundreds who have been rescued, to see whole communities growing up of men and women whose lives have been saved, and who are engaged in saving the lives of others. I have lived to see the spread of the new light from nation to nation until it has enriched the globe and finds expression today in the gathering of the International Congress of Tuberculosis, with all that it means to science, philanthropy, and the brotherhood of men." To this we may add Myers's comment in 1927: "Perhaps nothing in the whole field of medicine, in the past fifty years in America, has done so much directly or indirectly to relieve suffering and extend the years of usefulness of so many people as the principle which Dr. Trudeau laid down at Saranac Lake."

Francis B. Trudeau, writing an introduction to the third edition of his father's autobiography in 1943, stated that: ". . . though the specific cure for tuberculosis still eludes us, the good work goes on [and] the two major interests of his life, the sanatorium care for the tuberculous, and the scientific research into the origin and nature of the disease, which had such modest beginnings in the 'Little Red' and the homemade laboratory, have developed in ways undreamed of a generation ago." The status of treatment remained as it was expressed in the inscription on the statue of the great American pioneer in the treatment of tuberculosis:

> *Guérir quelquefois,*   (To cure sometimes,
> *Soulager souvent,*   To relieve often,
> *Consoler toujours.*   To comfort always.)

The same noble but unavoidably discouraging thought is echoed in the work of Brown, who wrote in 1941: "We are still today in the period of the sanatorium, which might be said to have begun about the middle of the 19th century. But this last division needs subdivision and it is suggested that the years from 1890 to 1900 be termed the rest period; that the time from 1900 to 1915 be known as the period of roentgenology; and even though it too is still with us, the period from 1915 to the present day should be called the period of surgical treatment of pulmonary tuberculosis."

Congresses and Societies for Prevention of Tuberculosis

Biggs, in 1893, suggested procedures for the control of tuberculosis which have probably been the basis of all modern control programs in the United States. Pulmonary tuberculosis was declared an infectious and communicable disease and reportable to the Sanitary Bureau. Emphasis was laid upon the isolation and treatment of persons ill with the disease and the building of sanatoria by states, cities, and counties.

Soon after Trudeau's sanatorium was established, the first society for the prevention of tuberculosis in this country was founded in Pennsylvania, in 1892. The founder, L. F. Flick, a consumptive himself, wrote: "With acceptance of the infectious nature of tuberculosis, especially after its demonstration through the discovery of the tubercle bacillus, men's minds naturally turned towards organization and cooperation for the spread of knowledge about tuberculosis and for the prevention of the disease. Apparently the first coordinated effort was made in France. Herard, president of the International Congress on Tuberculosis in Paris, in 1905, in his address, stated that an International Congress on Tuberculosis had taken place in Paris in 1867, on the occasion of the Universal Exposition . . . another Congress had been held in 1888 . . . a third in 1891, presided over by Villemin. Still two more had been held in Paris, one in 1893 . . . and one in 1898, presided over by Nocard."

Numerous groups of physicians and laymen in various countries organized associations for the prevention of consumption and other forms of tuberculosis. One such was formed in 1898 in London. Its purpose was to make known to the public the contagious nature of the disease and the preventive and sanitary measures that should be taken against it. In 1904 there were 38 voluntary tuberculosis societies in the United States, and in that year the National Association for the Study and Prevention of Tuberculosis (now the National Tuberculosis Association) was organized. Trudeau was chosen as the first president. A Christmas seal campaign was started by a Danish postmaster and was promoted in the United States by Jacob Riis, who himself had lost six brothers to tuberculosis. This helped to solve some of the financial problems of the new association. The association's activities were directed toward more sanatoria, enlarged public health campaigns, mass detection through chest X-rays, and other preventive measures.

At the International Congress in Paris in 1905, the delegation of the National Association invited the congress to hold its next meeting in the United States. According to Flick, "The entire United States entered heartily into the movement, and three years were spent in preparation for the congress. The meeting was held in Washington and there was a large attendance from all parts of the world. President Roosevelt was President of the Congress. The proceedings and transactions were published in seven volumes and distributed to all members of the congress. After the adjournment of the congress the International Society met in

Philadelphia. A wonderful impetus was given to the movement for the study and prevention of tuberculosis by these meetings."

One should also mention the International Union against Tuberculosis, with headquarters in Paris, which carries out a variety of activities, and has been organizing the International Tuberculosis Congresses since 1920.

The practical developments resulting from the discoveries of the late nineteenth and early twentieth centuries can be briefly formulated as: prophylaxis, surgery, and rest.

The failure of Koch's "tuberculin" did not prevent other investigators from searching in the tuberculosis organism itself for a preventive, if not a cure, for the dangerous enemy. Beginning with Friedman's turtle bacillus in 1910, various vaccines were tried unsuccessfully. Much better results were obtained from the prophylactic use of the BCG (Bacillus Calmette-Guerin), introduced in 1921.

Much progress was made during this period in the application of surgery to tuberculosis. The most important was the introduction of the artificial pneumothorax recommended by Forlanini in 1882, as pointed out in a previous chapter. Permanent collapse therapy by thoracoplasty was a popular form of therapy until recent years.

Finally, rest and collapse therapy, in special sanatoria, were considered by many as the only possible salvation for the tuberculous, recalling Shelley's beautiful *Lines Written Among the Euganean Hills:*

> . . . soft sunshine, and the sound
>   Of old forests echoing round,
>   And the light and smell divine
>   Of all the flowers that breathe and shine:
>   We may live so happy there,
>   That the spirits of the air,
>   Envying us, may even entice
>   To our healing paradise
>   The polluting multitude . . .

As much as four years of rest was considered necessary for the healing of pulmonary tuberculosis. "Unfortunately," as Webb has said (1936), "too many victims are inclined to agree with Homer, that too much rest itself becomes a pain, and cooperation is apt for fail."

In 1953, Long wrote: "In future socio-medical history the nineteenth century will probably be viewed as the century of tuberculosis. In the most, not to mention the least, civilized countries, tuberculosis was then by far the most destructive disease. To be sure, the nineteenth was not the century when tuberculosis was at its peak. Higher mortality rates prevailed before that. But previous centuries were characterized by enormous general death rates and wave after wave of highly fatal epidemics of acute disease. Tuberculosis in those centuries was accepted as a phenomenon inevitable in a large part of the population, as natural for some people as old age for others. It was common but inconspicuous."

This completes the story of tuberculosis until 1943. Then the picture looked gloomy indeed. In spite of the great progress made in discovering the nature of the disease, in methods of prevention, and in suitable care of the afflicted, the hopelessness of ever finding a proper cure was devastating. The solution came from quite an unexpected source, namely, the soil. Mother Earch brought the answer, thus justifying the ancient author of the book of Ecclesiasticus: "The Lord hath created medicines out of the earth; and he that is wise will not abhor them."

« Part III »

CHEMOTHERAPY, CONTROL, AND FINAL CONQUEST OF TUBERCULOSIS

« 6 »

The Advent of Antibiotics

One of the most interesting potential approaches to the tuberculosis problem was completely overlooked by experts in the fields of clinical tuberculosis and public-health control of the disease. That approach came about through the study of the lowly microbes by investigators who usually had little interest in or knowledge of the medical care of patients with tuberculosis. It has been known for many years that saprophytic microbes, the universal scavengers, inhabit in large numbers our soils and our seas, the air we breathe, and the water we drink. It was Pasteur who prophesied in 1867, just about the time that Villemin carried out his classical studies on the infectious nature of tuberculosis, that the time might come when we would fight disease-producing organisms with harmless or saprophytic microbes. It was also generally known that tuberculosis bacteria, like other microbes causing infectious diseases, do not survive long when introduced into soils and into natural bodies of water; this was the basis of my own work and that of many others on possible sources of materials that would be antagonistic to the tuberculosis organism.

The results of studies carried on in my laboratory in 1932 on the survival of the tuberculosis organism in soils and in water basins (p. 3) led to the following conclusions: (1) certain strains of *Mycobacterium tuberculosis* die out only very slowly when introduced into sewage and into ordinary potable water; (2) various microbes inhabiting the soil, notably certain protozoa, have the capacity to ingest large numbers of the tuberculosis organism; (3) *M. tuberculosis* is able to multiply in sterile soil, but it does not grow and is slowly destroyed in fresh or nonsterilized soil; (4) certain fungi are able to repress its development, especially in manured soil.

The above study was completed in 1935. I did nothing further to bring the above results to their logical conclusion. No attempt was made to examine the chemical mechanisms involved in the destruction of the tuberculosis organism under those particular conditions. My general impression at that time was that these results seemed to lead nowhere. It must be mentioned here that similar observations were made and conclusions reached in various other laboratories throughout the world. In the scientific climate of the time, the results did not suggest any practical applications for the treatment of tuberculosis. Another important reason for not following this problem further was that I lacked the necessary facilities to attack it on a larger scale. My laboratory was rather small and was devoted primarily to the study of the microbial population of the soil and its role in soil processes. Pathogenic microbes, especially those that required extensive animal experimentation, were at that time somewhat out of my field of investigation.

There were other reasons why the problem was not followed further at that time in my own or in other laboratories. Available funds were limited for undertaking an extensive study of the relationship of the microbiological population of the soil and pathogenic organisms that find their way into the soil. The government agencies sponsoring the study did not appear to be interested in supporting it further. My own experience in working with acid-fast bacteria in general and the tuberculosis organism in particular was limited. I decided, therefore, to limit my efforts to the study of the effect of antimicrobial agents upon the common gram-positive and gram-negative bacteria that are responsible for many of the infectious diseases of man and animals.

The possibility of attacking tuberculosis through antimicrobial agents seemed at that time rather remote. Several other laboratories were engaged in studies of the chemical composition of the tuberculosis bacterium; this organism was found to have a rather heavy coating of waxy material, which was difficult to digest by ordinary enzymatic systems. Although various suggestions were made that certain special enzymes might be found that could attack this material and thus become of practical value in the treatment of the disease, the idea did not seem to have any practical potentialities. Eight more years were to pass before I became aware of the great potentialities of this new type of approach, involving a search for saprophytic organisms capable of inhibiting the growth of the *M. tuberculosis* both in the test tube and in the living body.

I shall never forget the bright morning of June 1, 1943, when a small group of people met in one of the rooms at the Pennsylvania Hotel in New York to discuss possible approaches to the problem of the chemotherapy of tuberculosis. Those present were Leroy U. Gardner, the bacteriologist of the Saranac Laboratory; William C. White, of the National Tuberculosis Association; several representatives from pharmaceutical companies interested in developing chemotherapeutic agents against infectious diseases; and two or three others from different university laboratories, including myself. No definite program had been prepared for this meeting; the general discussion had a discouraging undertone, and it did not lead to any particular conclusions. The sulfa drugs, which had been available for a few years, were effective against many infectious diseases, but seemed to have little or no effect against tuberculosis. Penicillin, which had recently made its appearance and was beginning to find a place in the treatment of various infections produced largely by cocci and certain gram-positive bacteria, was likewise ineffective against tuberculosis. Except for the sulfones, which had attracted recent attention, no synthetic compounds were known to have any significant effect upon this disease.

What was to be done now? Could one suggest new approaches that seemed to be promising and should be investigated further? White spoke about the possible isolation of digestive enzymes from earthworms. The National Tuberculosis Association, the Research Corporation, and Merck and Company were supporting a study by a scientific group in Washington in an attempt to determine the role of earthworms in digesting the tuberculosis organism. As a result of this study, the suggestion was made that "whatever causes this seeming destruction of the bacilli in the earthworm, without injuring it, may be an enzyme, a product which will interfere with the ordinary function of some other body, in this case, the tubercle bacillus" (Nicolson, 1943).

My own reaction to this report was definitely negative. I expressed sincere doubt whether such an approach, through the study of the digestive mechanisms in an animal body, would work at all, and even whether further investigations in this direction were justified. I said: "After all, the bacteria causing tuberculosis are to be destroyed not in the test tube, but in the human body, and any enzyme system powerful enough to bring this about would certainly digest the human organs as well."

White was not very happy about my comments, and spoke up rather angrily: "How do you propose to go about this problem?"

My answer was simple: "The antibiotics will do it. Just give us time. Sooner or later, we are bound to find one or more chemical agents that will be able to bring this about. They will kill the bacterium not by digesting it, but by interfering with its metabolism and its growth, without injuring the host. After all this would be based upon the principle of chemotherapy, as enunciated by Paul Ehrlich. Let us hope that such antibiotics will be found, and that they will not be too toxic to the human body."

My argument did not carry much weight. It was discussed, and was finally dismissed as purely speculative. To me, however, these words expressed not merely hope and speculation. I was well aware, from a general study of the literature, of the potentialities of this type of approach; my own work at that time, on the production of antibiotics by actinomycetes, had already yielded several chemical agents, notably actinomycin and streptothricin, both highly interesting antibiotics, active against a great number of different bacteria. Although these had not yet found a place in general chemotherapy because of their toxicity, and were not very active upon the *M. tuberculosis,* they pointed to a new approach to the problem of chemotherapy (fig. 12).

The first observation of the effect of saprophytic microbes upon

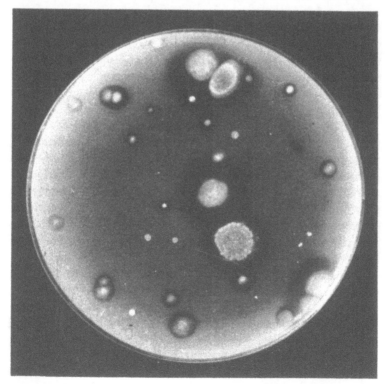

FIG. 12. A sample of soil plated out on an agar medium previously inoculated with a bacterial culture, showing colonies of bacteria, actinomycetes, and fungi, some of which produce a growth-inhibiting effect upon the bacterial culture.

the tuberculosis organism had been made within three years after Koch presented his famous paper. In 1885, Cantani, an Italian bacteriologist, reported that the virulence of tuberculosis in human corpses disappeared under the influence of putrefactive processes. He conceived the idea of treating pulmonary tuberculosis by having the patient inhale cultures of nonpathogenic bacteria. A common saprophytic organism known as *Bacterium termo* was used for this purpose. A 42-year-old woman who had a large cavity in the superior lobe of the left lung, who suffered from fever and coughing, and whose sputum test was positive, was treated by inhalation of a liquefied gelatin culture of the saprophyte diluted with meat broth. The pathogenic bacteria in the sputum were rapidly reduced in numbers, and eventually they disappeared completely; the fever diminished, and the patient's condition showed definite improvement. One case, of course, was not impressive. Spontaneous recoveries were not infrequent.

This observation did not remain completely isolated. Three years later, in 1888, Babés, a Rumanian investigator, demonstrated that products of various saprophytic bacteria, comprising both gram-positive and gram-negative forms, such as staphylococci and the prodigiosus organism, have the capacity of inhabiting the growth of *M. tuberculosis;* streptococci and pneumococci, however, seemed to be favorable to its development. Babés noted the importance of associative organisms in influencing the course of tuberculosis.

Another Italian bacteriologist, Nannotti, reported in 1893 that tubercular inflammation of the knee joint tended to subside as a result of the spontaneous occurrence of an erysipelas infection. The use of streptococci or their metabolic products, filtered and sterilized, resulted in a modification of the tuberculous process; abscesses grew smaller and fistulas tended to heal.

The results of these studies, however, had not been universally accepted. The evidence of a specific antagonism of certain bacteria toward the tuberculosis organism was questioned, and the reported results were explained as possibly attributable to simple inflammation caused by the accompanying bacteria.

New observations had continued to appear in the literature, tending to confirm the earlier reports of Cantani and Babés. Various saprophytic bacteria were found to be capable of inhibiting the growth of the tuberculosis organism; even the excretion products of these saprophytic bacteria exerted the same effect, but to a lesser degree. When a drop of a culture of one of the above bacteria was added to a culture of *M. tuberculosis,* growth was stopped. The inoculation of the saprophytes into experimental animals was found to prevent the growth of the tuberculosis organism.

It became soon apparent that the ability to produce tuberculostatic and tuberculocidal agents is widespread among microbes, including not only bacteria, but also actinomycetes and fungi. Unfortunately, the significance of these results was not fully recognized. Another decade or two were to elapse before real progress was to be made in this direction. The early studies on the antagonistic effects of various microbes and their metabolic products on the growth of *Mycobacterium tuberculosis* in culture and on the course of the disease produced by this organism were then followed by more carefully planned investigations.

Rappin reported in 1912 that cultures of a number of spore-forming bacteria, such as *Bacillus subtilis, B. mesentericus,* and *B. megatherium,* have a suppressive effect on the development of the tuberculosis organism. When filtrates of these spore formers were injected into rabbits infected with the *M. tuberculosis,* the development of the disease was prevented and no toxic manifestations occurred. This specific antituberculous activity was believed to be attributable to antibacterial substances produced by the spore-forming bacteria. Unfortunately, Rappin designated these substances as diastases, thus suggesting their enzymatic nature. Although the actual results obtained were highly significant, the misnomer of the chemical agents involved in the process led to confusion and, therefore, the results failed to arouse sufficient interest among pharmacologists and clinicians. They were not followed up properly by other studies along these lines.

Vaudremer demonstrated in 1913 that a culture of the tuberculosis organism lost its pathogenic power when brought in contact for 24 hours at 39°C. with a filtrate of the fungus *Aspergillus fumigatus.* When the culture thus treated was injected into rabbits, it seemed to vaccinate the animals against tuberculosis, since the injection, a month later, of fresh and virulent cultures of *M. tuberculosis* did not cause any infection comparable to that of the controls.

Vaudremer utilized the extracts of this fungus for the treatment of some two hundred patients suffering from tuberculosis. The fungal preparations seemed to be nontoxic. Although favorable results were obtained in some cases, quite unexpected results were observed in others. Vaudremer concluded that any generalizations based on these studies would have to be considered as premature, since they hardly justified any recommendation that this method of treatment of tuberculosis be further extended. This was expressed by Vaudremer quite emphatically: "It is still too early to draw any conclusions with regard to the treatment of tuberculosis in man. Since 1910 we have treated more than two hundred patients with extracts of *A. fumigatus;* the treatment was practiced in several hospitals and sanatoriums in Paris. From the facts observed one may conclude that these injections are innocuous, since they never caused a febrile reaction. Occasionally one even observed an unexpected recovery. In other cases one may observe a temporary improvement; but unfortunately the instances are still numerous in which tuberculosis takes its course."

These results tended, however, to establish the fact that the *M. tuberculosis* is highly sensitive to the products of various saprophytic bacteria and fungi. The living microbes themselves, and their products as well, were thus shown to have the capacity of inhibiting and even destroying the tuberculosis organism, both in the test tube and in the living body. Again, unfortunately, no attempt was made to utilize these observations for the purpose of isolating a practical chemical compound for the treatment of tuberculosis.

In 1940 the Italian bacteriologist Zorzoli, working at the famous tuberculosis institute "Carlo Forlanini," demonstrated that culture filtrates of certain species of fungi, especially those belonging to the *Mycotorula* and *Aspergillus* groups, exert a marked antibacterial effect on both human and bovine forms of M. *tuberculosis*. This effect was found to take place not only *in vitro,* or in test tube cultures, but also *in vivo,* namely, in inoculated pigs.

My own interests were centered at that time upon the study of a certain group of saprophytic microbes, known as actinomycetes (fig. 13), and their ability to inhibit the growth of various pathogenic and nonpathogenic bacteria and fungi. I was led to these investigations by my previous studies, dating back a quarter of a century, of the microbiological population of the soil, especially the fungi and actinomycetes. The latter appeared to exert very peculiar growth-depressive effects upon the soil bacteria. Others, as well, had observed this phenomenon, but very little work had been done on the possible sensitivity of the tuberculosis organism to the effect of actinomycetes.

After various preliminary studies of the production of antibiotics by actinomycetes which resulted in the isolation of actinomycin and streptothricin, actual studies of the effect of antibiotics upon the tuberculosis organism were begun in 1943. Several significant facts were soon demonstrated from which various conclusions could be derived. These were summarized later (Waksman, 1947) as follows:

> 1. Certain species of actinomycetes and especially certain strains are capable of exerting an antituberculosis effect.
> 2. The antibiotics produced by these organisms vary greatly in their chemical composition and in their toxicity to animals.
> 3. The growth-inhibiting effect of these antibiotics on M. *tuberculosis* and on other bacteria varies greatly. Their antimicrobial spectra range from extremely narow ones, comprising activity only against certain mycobacteria, to exceedingly wide spectra, being active against not only numerous grampositive and gram-negative bacteria but also fungi.

Some of these facts were in my mind when I made my remarks to William C. White on that memorable morning in June, 1943. We were already quite familiar at that time with the great poten-

Fig. 13. A group of cultures of actinomycetes isolated from the soil, belonging to the genus *Streptomyces*.

tialities of certain antibiotics, notably actinomycin and strepto-thricin, that we had isolated in our laboratory, as well as with tyrothricin and penicillin, isolated in other laboratories. Unfortunately, none of these compounds was known at that time to be active against the tuberculosis organism.

The discovery of streptomycin was a result of a comprehensive screening program of various groups and species of soil-inhabiting microbes that had the capacity to inhibit the growth of different bacteria, notably those that could cause infectious diseases in the human or animal body. The cultures of microbes that showed such inhibiting activity were grown in artificial media, and by means of special chemical procedures the active substances produced by these microbes were isolated, then concentrated and purified. These chemical substances, designated as "antibiotics," were then tested for their activity against different bacteria, including pathogenic forms. Their effect was found to vary greatly, depending on the chemical and biological characteristics of the substances and the specific nature of the organisms used for testing purposes.

After certain preliminary studies on the antagonistic interrelations among soil microbes that were carried out in my laboratory for nearly a decade, I began in 1939 a systematic search for various saprophytic microbes, largely those found in the soil or in the dust, that had the capacity to prevent the growth of other microbes, notably disease-producing bacteria. In this work, I had the help of a number of graduate students and assistants, without whose collaboration such extensive studies would never have been possible. It is sufficient to mention H. B. Woodruff, H. Robinson, Elizabeth S. Horning, Doris Jones, A. Schatz, H. Christine Reilly, Elizabeth Bugie, D. B. Johnstone, O. E. Graessle, and Dorothy R. Smith.

I paid particular attention to one group of microbes occurring extensively in soil, namely the actinomycetes. These filamentous branching bacteria are frequently spoken of, quite incorrectly, as "soil molds." There were two important reasons why I used largely these microbes as potential producers of antibiotics:

1. I had been familiar with the nature and activities of actinomycetes for nearly a quarter of a century, or since 1914, when I first undertook experiments on the occurrence and biochemical activities of soil microbes at Rutgers as an undergraduate and then as a graduate student, and later (in 1916) at the University of California, where I obtained my Ph.D. degree.

2. By a sheer stroke of good fortune, the actinomycetes proved to be highly productive of antibiotic substances. They have since attracted attention all over the world, because of this particular property.

The first antibiotic that we isolated from a culture of an actinomycete in 1940 was designated as *actinomycin*. The organism producing this substance was described as *Actinomyces antibioticus*. The substance was a red-pigmented compound; it was highly active against various bacteria, especially the gram-positive forms. Unfortunately, when injected into experimental animals, it proved to be extremely toxic; it could hardly be considered a potential therapeutic agent. (More recently, in 1954, another form of this compound, designated as actinomycin D, was isolated in my laboratory; it is now recognized for its specific effect upon certain tumors and has come to occupy an important place in the treatment of such forms of cancer.)

The search for new antibiotics continued, and two years later we succeeded in isolating, from another actinomycete culture, a new and distinct type of antibiotic, designated as *streptothricin*. This was a water-soluble white powder, which was active against a variety of gram-positive and gram-negative bacteria, as well as certain fungi. At that time we did not test its action upon the tuberculosis organism; however, when this was done later by Woodruff and Foster (1944), streptothricin was found to be active against this bacterium as well. It was not so toxic to experimental animals as was actinomycin, but it still possessed a certain degree of delayed toxicity, so that the animals cured from artificially produced infections later died.

Nevertheless, streptothricin represented a very important step in our progressive search for new antibiotics. In the process of its isolation, several methods and approaches toward the discovery of more desirable kinds of antibiotics were developed. Our screening program continued uninterruptedly. New cultures of different microbes were isolated. These were grown in various media and tested by several new procedures for their effects upon various bacteria. In September, 1943, less than a year after the isolation of streptothricin, a new compound, similar in its chemistry and in certain biological properties, but less toxic to experimental animals, was isolated and designated as *streptomycin* (Schatz, Bugie, and Waksman, 1944; Waksman, Bugie, and Schatz, 1944). Its growth-inhibiting effects upon the tuberculosis organism was soon established (Schatz and Waksman, 1944).

I have often been asked who really isolated the streptomycin-producing culture from the soil. My usual reply has been in the form of a story: One day early in August, 1943, a New Jersey farmer observed that one of his chickens seemed to be suffering from a peculiar ailment that affected its breathing. Fearing an epidemic of some kind, the farmer took the chicken immediately to the poultry pathologist at the nearby Agricultural Experiment Station. As is usually done in such circumstances, the pathologist used a small cotton swab to transfer the contents of the chicken's throat to several culture plates. After incubating the plates for several days, he observed three colonies of actinomycetes developing on one of the agar plates. At that time one of my assistants was working in this laboratory, studying virus techniques. The poultry pathologist gave her the plate, saying, "You had better take this to Dr. Waksman. I know that he is interested in actinomycetes—perhaps he might like to see whether these cultures are of any interest to him." One of my graduate students was working at that time on the screening of actinomycetes for antibiotic-producing organisms. When the plate arrived, I handed it to this student and asked him to transfer the colonies to fresh agar slants, from which cultures could be made in order to determine their antibiotic-producing capabilities. The techniques for doing this type of cultivating and testing had been worked out in my laboratory during the previous four years.

When the cultures were isolated and cultivated on agar slants, they were brought to me. I identified at once the organism growing on the agar as one that had been isolated from the soil in the same laboratory by Curtis and me (Waksman and Curtis, 1916) some twenty-eight years previously. We had given it at that time the name *Actinomyces griseus.* Since the generic name of this group of organisms was recently changed by S. A. Waksman and A. T. Henrici (1943) from *Actinomyces* to *Streptomyces,* the name of the freshly isolated culture would now be *Streptomyces griseus.*

The antibiotic activity of the new culture seemed at first to be similar to that of streptothricin. I then asked another student to start several cultures from the agar, using different broth media, so that an attempt could be made to concentrate the active material and to determine its nature. Methods for these procedures had been developed by Woodruff and me (Waksman and Woodruff, 1942) when we were isolating streptothricin. As soon as the active concentrates of the new culture were ready, I asked two other assistants, aided by students and members of the animal husbandry group at the experiment station, to test their effectiveness in experimental animals. Although the type of activity of the new preparation was found to be similar to that of streptothricin, there was one very important difference. It was less toxic. We named the new substance streptomycin. Five months later, in January, 1944, this discovery was announced in a paper by Schatz, Bugie, and Waksman.

By one of those peculiar happenstances which seem to lend credence to the precognition concept of the parapsychologists, the name for the new antibiotic was already at hand. Seven months before, I had announced to the students in my laboratory that A. T. Henrici and I had decided to establish a new genus of aerobic actinomycetes to be called *Streptomyces.* Prophetically, one of the students went at that time a step further with the observation—"Now we have a new name for a new antibiotic, streptomycin."

Many important things happened before the chicken came into the picture, however, and it is, therefore, difficult for me to say just exactly when the streptomycin story had its beginning. My studies on the actinomycetes which were basic to the later discoveries go back to 1914. My studies of antibiotics were started in 1938.

One could argue that streptomycin was on its way in 1915 when I first learned to recognize the nature and physiology of the organism we called *Actinomyces griseus*. Other important events along the way were the recognition of the actinomycetes as producers of antibiotics with the isolation of actinomycin in 1940, and our development of procedures in 1941-1942 leading to the isolation of streptothricin.

If you ask who was responsible for the isolation of the particular streptomycin-producing strain of *Streptomyces griseus,* the answer must be that it was the chicken, because it was she who picked up the culture from the soil. But if you ask who discovered streptomycin, the decision is more complex.

Was the discovery made by the poultry pathologist who recognized the actinomycete colonies on the culture plate inoculated from the chicken's throat?

Was it made by the student who worked under my direction and used the methods that I had developed for the further cultivation and testing of these colonies?

Was it made by my students and me in my laboratory or by my students working in the Merck laboratories, when we demonstrated the control of several experimental diseases in animals by streptomycin and its limited toxicity?

Was the discovery made when William H. Feldman and H. Corwin Hinshaw of the Mayo Clinic first tested and soon reported the activity of streptomycin in the control of experimental tuberculosis? Or was it made when the first clinical cases treated with streptomycin were found to benefit from it?

When did the actual discovery of streptomycin and its effectiveness upon tuberculosis first come about?

You might just as well ask the artist who creates a beautiful mosaic, "When was the real creation made? Did it begin with the artist's conception of the design? Did it begin with the laying of the first stone or of the last stone? Did it begin with the central figure, or with the signature of the artist after the work was completed?"

An interesting sidelight further emphasizes the difficulties in trying to specify origins. In 1950, at my request, A. Kelner at Harvard University irradiated the 1915 culture of *Actinomyces griseus* that was kept alive in my culture collection, and obtained a mutant that produced streptomycin. Does that signify that the 1915 culture, isolated by Waksman and Curtis, had at that time the capacity to produce streptomycin and lost it as a result of continued cultivation on artificial media, or that it gained the property always present in a weak state, upon irradiation?

On November 1, 1945, I submitted a statement reading as follows (Waksman, 1945) to a subcommittee of the Committee on Military Affairs of the United States Senate:

"The discovery of streptomycin is another illustration of the importance of planning biological research in the solution of a given scientific problem. When a comprehensive project of a systematic study of the production of antibiotic substances by different groups of microorganisms was undertaken at the New Jersey Station, various approaches to the problem were at first examined. It did not take long to discover that microorganisms capable of producing antibiotics are very abundant in the soils and in composts, that such microorganisms produce more than one type of antibiotic and that these substances vary greatly in chemical nature, selective antibacterial activities, toxicity to animals, and activity in the animal body. It also became soon apparent that such a study would involve detailed surveys of large numbers of organisms and that long tedious research is involved . . . With limited funds and still more limited personnel, this was not a very easy problem. Fortunately, one of the foundations and one of the industrial organizations came to the support of this project. The subject attracted graduate students to the University, intent upon making this problem a subject for their doctor's thesis. A dozen or more collaborators were thus engaged in this comprehensive investigation."

«7»

Streptomycin and the Beginnings of True Chemotherapy of Tuberculosis

When streptomycin was first tested for its action upon various pathogenic bacteria, including *Mycobacterium tuberculosis,* it was found to be highly effective in bringing about rapid inhibition of their growth on artificial agar media. It was also found to be highly active against bacteria injected into animals; infection was prevented and complete cures were brought about after infection. Clinical studies on streptomycin were now undertaken. It was soon established that this antibiotic possessed many of the desirable properties of a potential chemotherapeutic agent in the treatment of numerous infectious diseases, including tuberculosis.

In the discovery and in the testing of the effectiveness of streptomycin, certain dates are important because of their historical significance. The first of these dates is August 20, 1943, when two cultures of an organism, later recognized as *Streptomyces griseus* (Waksman and Henrici), a species of actinomycete, were isolated from the swab of a chicken's throat. These two cultures, varying somewhat from each other, produced the new type of antibiotic, streptomycin (figs. 14, 15). Later it was found that

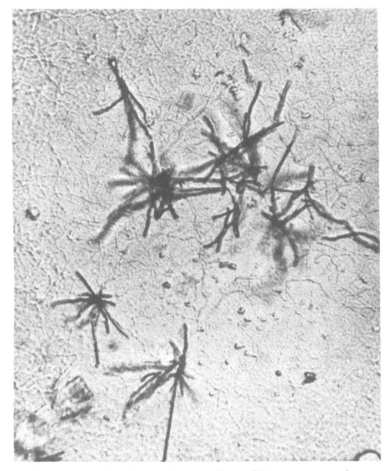

FIG. 14. Aerial and vegetative mycelium of *Streptomyces griseus*.

other cultures of the same species and even of other species could be isolated from various soils and were capable of producing the same type of antibiotic or certain chemical modifications of it.

The isolation of the new antibiotic was announced in January, 1944. Because of prevailing war conditions, the publication in which this announcement was made did not appear until Febru-

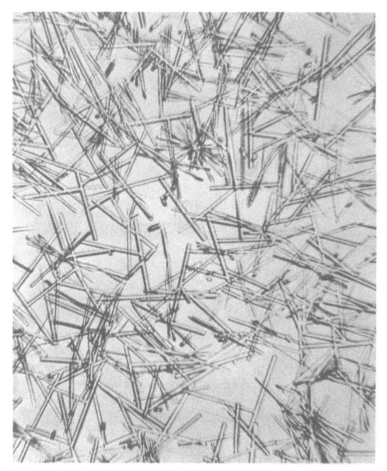

FIG. 15. Crystals of streptomycin. (Courtesy of Merck and Company.)

ary. Two months later, we reported further in another publication (Schatz and Waksman, 1944) that this antibiotic was highly effective against *Mycobacterium tuberculosis* and other mycobacteria. We showed that it possessed decided tuberculostatic and tuberculocidal properties. The human strain of *M. tuberculosis* H 37 was particularly sensitive to streptomycin, the growth of the

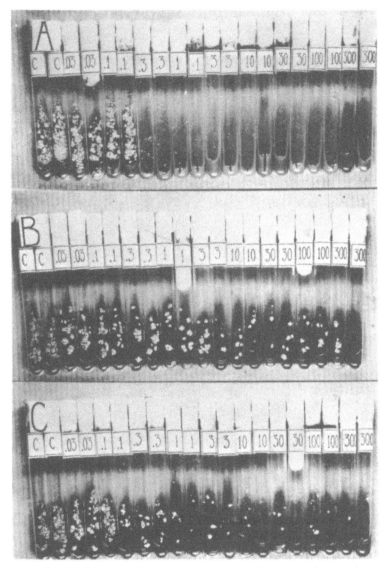

FIG. 16. Demonstration of effect of streptomycin on tuberculosis organism—the bacteriostatic and bactericidal action of streptomycin upon a human strain (H 37) of *M. tuberculosis*. The figures on the tubes give units of streptomycin, C being the control. *A.* Bacteriostatic action. *B.* Bactericidal action, three days. *C.* Bactericidal action, **seven** days. (AFTER SCHATZ AND WAKSMAN, 1944.)

organism being inhibited by only 0.2 microgram of the antibiotic (fig. 16). The tuberculosis organism was six or more times
as sensitive to streptomycin as were common aerobic bacteria,
such as *Escherichia coli*. The new compound was active upon
bacteria both *in vitro* and *in vivo*, and had only limited toxicity
to experimental animals.

First Animal Experiments Against Tuberculosis

The announcement of the effectiveness of streptomycin
upon the tuberculosis organism immediately attracted the attention of Dr. William H. Feldman and Dr. H. Corwin Hinshaw,
of the Mayo Clinic, who were interested at that time in finding
antituberculosis agents for testing in experimental animals. Shortly after reading our report, they came to see me to arrange for
testing the new antibiotic in experimental animals. Meanwhile, in
our own laboratories we had tested the effect of streptomycin
upon various common disease-producing bacteria inoculated into
animals, and had observed that all the infected animals were
saved. The following pathogenic bacteria were used in the test:
*Salmonella schottmülleri, Pseudomonas aeruginosa, Shigella gal-
linarum, Brucella abortus,* and *Proteus vulgaris.* The results thus
seemed to be highly encouraging (fig. 17).

On the basis of our studies, I felt justified in promising to supply the Mayo Clinic investigators at once with some of our crude
preparation of streptomycin, which at that time contained only
5.8 per cent of the active substance. I also arranged with Merck
and Company—manufacturers of chemicals and pharmaceuti-
cals, located only a few miles from our laboratory, with whom I
was collaborating in the purification and testing of our newly
isolated antibiotics—to produce a large quantity of streptomycin
which would be required for more extensive trials. In their first
experiment on tuberculosis in guinea pigs, Feldman and Hinshaw soon demonstrated the high effectiveness of this antibiotic. The striking antibacterial activity of streptomycin against

CONTROLS

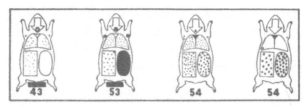

TREATED

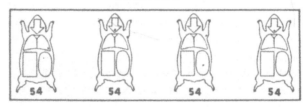

FIG. 17. First preliminary animal experiment on the effectiveness of streptomycin against tuberculosis. Amount of tuberculosis noted grossly in treated and untreated animals, shown schematically, the severity of infection being indicated by the depth of dark color. The numerals give the length of life in days after infection. The bottom black bar indicates that the animal died. (AFTER FELDMAN, HINSHAW, AND MANN, 1945.)

tuberculosis *in vivo* was thus established. The treatment of the infected animals with streptomycin resulted in the resolution, fibrosis, or calcification of lesions and in the apparent eradication of the infection in at least 30 per cent of the animals; positive reactions to tuberculin tests became negative, indicating that the animals were free from tuberculosis.

These preliminary studies encouraged the Mayo Clinic investigators to visit me again, on July 10, 1944, to discuss the results of the first experiments and to plan more extensive animal studies. They even began to foresee the immediate clinical treatment of tuberculosis in man. Merck and Company at once agreed to prepare additional quantities of streptomycin. All of us concerned in these investigations were of the opinion, however, that the preliminary results must be kept strictly confidential, so as not to arouse undue or premature public optimism. This is confirmed in a letter written to me by Dr. Feldman a week later, on July 19:

"You will be interested to know that the guinea pigs received their infecting dose of tubercle bacilli yesterday morning and their first dose of the respective medicants, streptothricin and streptomycin, at 6:00 last evening. The dose schedule calls for an injection every 6 hours. So far we have been unable to obtain any additional laboratory assistance so Dr. Hinshaw and I have taken over the task of administering the substances. We felt that if we were to wait until we got help the program might be delayed indefinitely. We consider knowledge of streptothricin and streptomycin as privileged information, so much in fact, that we do not use the terms in the laboratory but refer to them as A_1 and A_2. I agree with you that it would be very unwise to have the information reach the public prematurely." (See fig. 18.)

Ten years afterward, looking back upon these early experiments on the effect of streptomycin, Feldman (1954) had this to say:

"The results of the long-term experiment were impressively favorable. The findings demonstrated the unquestionable ability of streptomycin to reverse the potentially lethal course of well-established inoculation tuberculosis in guinea pigs, and the relatively low toxicity and corresponding safety of purified streptomycin. These observations, Dr. Hinshaw and I believed, satisfied the prerequisites of a therapeutic substance worthy of trial in clinical tuberculosis. As a result of these additional investigations it was established (1) that bovine as well as human tubercle bacilli are sensitive to the action of streptomycin and that the action is not strain-specific; (2) that it is not necessary to administer the drug at frequent intervals every few hours to obtain satisfactory therapeutic effects; (3) that streptomycin is not effective in suppressing experimental infections due to tubercle bacilli resistant to the drug *in vitro;* and

CONTROLS

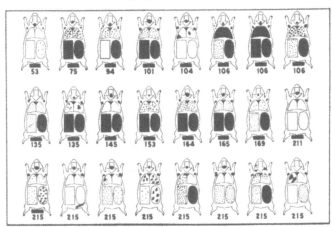

STREPTOMYCIN SERIES
TREATED AFTER 49 DAYS

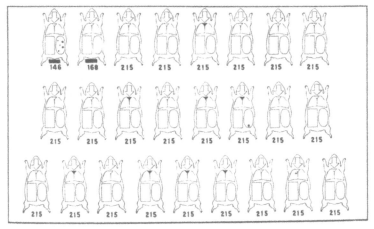

Fig. 18. More extensive series of animal experiments on effectiveness of streptomycin upon tuberculosis. (AFTER FELDMAN, HINSHAW, AND MANN, 1945.)

(4) that the potency of streptomycin as an antituberculous agent is sufficient to modify favorably and often dramatically the progression of a tuberculous infection even though the animals are inoculated with large amounts of tubercle bacilli intravenously."

Although Feldman and Hinshaw were very enthusiastic about the efficacy of streptomycin in experimental tuberculosis, they used a very restrained tone publicly. Gradually, larger quantities of the drug were produced and placed at their disposal so that they could proceed with its clinical evaluation.

CLINICAL TREATMENT OF TUBERCULOSIS

The first clinical treatments of tuberculosis with streptomycin were carried out at the Mayo Clinic in the winter of 1944/45 (Pfuetze et al., 1955). The treatment of the first case is presented here verbatim, because of its historical significance:

"November 20, 1944, was the day on which streptomycin was first administered to a human being for the treatment of tuberculosis. The patient was a 21-year-old white girl who had progressive, far advanced pulmonary tuberculosis . . . When she was admitted to the Sanatorium in July, 1943, this patient had far advanced pulmonary tuberculosis involving the right upper lobe. During her first year of hospitalization she showed some improvement, but shortly afterward she began to have chills, fever, night sweats, and increased cough; a chest roentgenogram obtained in October, 1944, showed a pronounced increase in infiltration and cavitation in the right lung . . . At this point it was decided that the patient's progressively unfavorable course made the use of streptomycin justifiable and desirable. Accordingly, some of the material then being used in the guinea pig experiment was appropriated for this first clinical trial. As the therapeutic and toxic doses of streptomycin for human beings were entirely unknown, the drug was administered with extreme caution. At first the total daily dose was only 0.1 gm., and this amount was divided into eight doses given at three-hour intervals around the clock. From the first the drug was given by deep intramuscular injection. The first preparations of streptomycin were relatively crude . . . Between November 20, 1944, and April 7, 1945, the patient received five courses of streptomycin, each of which lasted ten to eighteen days. The treatment was interrupted partly to give her relief from side effects, but mostly because at that time a steady supply of streptomycin was not available."

This patient was discharged from the sanatorium on July 13, 1947, with a diagnosis of apparently arrested pulmonary tuberculosis. There was no deterioration in her condition as revealed by periodic examinations. She has since married and is the mother of three children, born in 1950, 1952, and 1954. In reporting her case later, Pfuetze *et al.* (1955) further remarked: "Inadequate as her chemotherapy was by present-day standards, sufficient evidence now exists to allow the conclusion that the streptomycin which this patient received in the first clinical trial of the drug was definitely beneficial and may even have been life-saving. It is a happy ending which has been duplicated tens of thousands of times in the streptomycin story."

The production and evaluation of streptomycin were now developing at a rapid pace. There was an ever greater and greater need for larger and still larger quantities of the drug. Chester Keefer, of the National Research Council, was requested by Hinshaw on Janury 23, 1945, or within one year after the first public announcement of the discovery of streptomycin, to see that everything possible was done to expedite the production of streptomycin in quantities sufficient for adequate clinical tests. The proper judicious rationing of the drug was urgently recommended by Keefer's committee.

Experimental studies on the effectiveness of streptomycin were proceeding rapidly. On January 29, 1945, a telegram reached me from Feldman, stating: "Long term crucial experiment streptomycin terminated today. Incomplete results indicate impressive therapeutic effects." Hinshaw was maintaining his two patients, one of whom was the 21-year old girl mentioned previously, on streptomycin. On March 9 he wrote: "The results are sufficiently encouraging to be tantalizing, and we cannot avoid the feeling that if we could give a million or more units (micrograms) a day we might have something more impressive."

On August 25 Hinshaw was able to report to me: "We have now treated a total of thirty-three patients, including those who are now under hospitalization, and continue to be quite optimistic. Dr. Esmond Long and Dr. Henry Sweany spent an entire day going over every single case that has received streptomycin. They expressed an unqualified opinion that early extensive lesions of tuberculosis of the so-called hematogenous type are reversed to an unmistakable degree. As a result of their survey they are planning to undertake the treatment of tuberculosis in the Army with streptomycin, setting up a group at Fitzsimons General Hospital and at Burns General Hospital in the very near future."

On October 5 Feldman wrote: "Observations regarding the clinical application of streptomycin continue to be cause for enthusiasm. New information is constantly being assembled and of course we feel very fortunate in being permitted to continue our studies in spite of the fact that the War Production Board has taken over the allocation of streptomycin. We feel very important, indeed, that the authorities saw fit to permit our work to continue in spite of the fact that the supply for other civilian uses has been curtailed or perhaps cut off."

Finally, Hinshaw stated in a letter to me on December 11: "I believe that we now have undeniable clinical and pathological evidence that streptomycin does modify the course of tuberculosis in man in a favorable direction, but the magnitude of this effect, its permanence and its applicability to the common varieties of pulmonary tuberculosis are matters which must be left undecided until the drug is produced in adequate quantity for long-term treatment."

Thus within two years after the discovery of streptomycin its importance in the treatment of various forms of tuberculosis was established. The most spectacular results were obtained in miliary tuberculosis, tuberculous meningitis, tuberculosis of the larynx, bone and joint tuberculosis, and pulmonary tuberculosis. In March, 1946, as a result of the treatment of the first 100 cases, the Mayo Clinic investigators felt justified in coming to rather optimistic conclusions. They summarized their results as follows (Hinshaw, Feldman, and Pfuetze, 1946):

"Streptomycin is an antibacterial agent which possesses the unique ability to inhibit the growth of *Mycobacterium tuberculosis in vivo,* both experimentally and clinically. The manifestations of tuberculosis may be suppressed both in experimental animals and in man, with at least temporary retardation of the pathologic processes as judged by objective criteria. In many instances streptomycin appears to suppress tuberculosis rather than to eradicate it, being apparently bacteriostatic rather than completely bactericidal in the concentrations which can be obtained in the tissues."

Tuberculous meningitis was definitely arrested by the use of streptomycin. The treatment consisted in daily intrathecal administration of 100 to 200 mg. of streptomycin for two to six weeks. In addition, patients received daily by intramuscular injection 2 to 3 gm. of the drug for six months. None of the patients who died in the first experiments had received intrathecal (intraspinal) therapy, without which improvement was frequently of a temporary nature, lasting only for a few weeks or months.

First Encouraging Results

These early cases were presented before the National Tuberculosis Association meeting in Buffalo in June of 1946. On June 12 I received a telegram from Feldman and Hinshaw which read: "Our streptomycin studies reported at National Tuberculosis Association were fully confirmed experimentally and clinically, establishing this as first effective chemotherapeutic remedy for tuberculosis. Hearty congratulations."

Hinshaw (1954) described this particular meeting and its consequences as follows:

"Our most dramatic emotional experience was at the meeting of the National Tuberculosis Association . . . when we presented our first clinical report on streptomycin to a national audience. We feared that disbelief of the results would be expressed because of the many previous false hopes raised by other 'cures' throughout all previous medical history . . . immediately after the Buffalo meeting, a conference with military and veteran's authorities was held in Washington, D. C. At this conference we presented our data, and the plans were formulated for expansion of the study." The most extensive study of a single drug in a single disease ever undertaken was now initiated. This was "made possible by a gift of a quantity of streptomycin by the manufacturers which was currently valued at one million dollars. Very soon the initiative passed to the Veterans Administration . . . This project involving hundreds of investigators and many thousands of patients has grown and prospered to become an unparalleled phenomenon in cooperative medical research."

Feldman (1954) said of this meeting:

"The next important event indicative of the increasing interest in streptomycin was the opportunity Dr. Hinshaw and I had to present a summary of our observations at the 42nd annual meeting of the National Tuberculosis Association . . . During the discussion that followed our presentations, we were completely surprised and greatly pleased when Dr. Walsh McDermott and Dr. Carl Muschenheim announced that they and their associates at Cornell University Medical College, New York, had confirmed our observations."

On September 7, the Committee on Chemotherapeutics of the National Research Council, under the chairmanship of Chester Keefer, reported the results of the first 1,000 cases of different infectious diseases treated with streptomycin. On December 12 the first Streptomycin Conference of the Veterans Administration was held in Chicago. Reporting on this Conference, Hinshaw wrote to me:

"The results are most consistent and precisely like those which we have secured here. As a result of this presentation, it has been arranged that nineteen hospitals in the Veterans Administration will be using streptomycin on all suitable cases immediately and before many weeks the drug will be made available for treatment of tuberculosis in all the remaining Veterans Administration hospitals. It begins to look as though practical application of the drug is somewhat ahead of our knowledge of many fundamental problems, but this is probably better than to have the problem lagging behind."

It was thus definitely established that chemotherapy of tuberculosis was possible, that the Great White Plague could be treated by means of drugs just as could numerous other infectious diseases of man and animals. Streptomycin pointed the way.

Two other early cases of treatment of tuberculous meningitis with streptomycin may be mentioned here. One case was reported by R. E. Cooke and his associates (1946) of the Yale University Hospital:

"M.M., aged one year, was admitted to the New Haven Hospital on May 12, 1945, with convulsions. The only other abnormal physical signs were dullness and suppression of breath sounds over the left upper chest anteriorly and posteriorly. Treatment consisted of streptomycin intramuscularly and intrathecally, beginning on the eighth hospital day (May 19), for a period of 30 days. Because of lack of the drug, treatment was interrupted almost seven days. Therapy was resumed on July 2 and continued for an additional 29 days. The total amount of streptomycin given by intramuscular administration was 75 million units. This therapy was carried out over a period of 68 days. The patient showed definitely marked improvement. On December 31, 1945 she learned to walk and gained weight consistently. Many neurological signs, including rigidity, convulsions, irritability, have disappeared. Spasticity of the right arm and leg was no longer present and the baby could now see. Her vision returned . . . Serial x-ray examinations of the chest were made at intervals of two weeks. After one month of hospitalization (June 14) there was clearing of the previously noted mottled densities throughout both lung fields. On September 24 (136th hospital day), the large mediastinal nodes could not be seen. There is no evidence of tuberculosis of the skull, vertebrae, long bones, or abdomen."

Another early case came to my personal attention about the middle of January, 1946. Louis Krafchik, a local practitioner in my own home town of New Brunswick, telephoned me one day that he had a patient, a 15-month-old boy, an only child, afflicted with tuberculous meningitis. The child had a history of fever, irritability, restlessness for six days, and a temperature of 103° F. He had previously been treated by another physician with penicillin and sulfadiazine, but without any effect whatsoever. When admitted to the hospital on January 21, the child appeared acutely ill; he was extremely restless and bit his hands continuously. Streptomycin therapy was initiated on January 24. About five weeks later the neurologic signs disappeared, and the child was able to stand in his crib and walk around with support. Because of exhaustion of the supply of the drug, therapy was discontinued 32 days after initiation. The temperature rose again and treatment was resumed. On March 27 the child was discharged. He walked with only little support; there was no rise in temperature and no neurologic abnormalities were evident. Five months after onset of illness, a complete clinical recovery seemed assured.

In an official report to the staff of a New Brunswick hospital, Krafchik (1948) described the successful treatment of a relapse of the disease in this patient: "November 21, 1946, eight months after discharge and 10 months after the original illness, patient had fever of 103°, he became irritable, lost appetite. Follicular tonsillitis and otitis media were found and he was given sulfadiazine and penicillin with good effect. He remained irritable, then developed nuchal rigidity and hyperactive reflexes. He was readmitted November 28, 1946. The patient was treated again intensively with streptomycin over a period of seven weeks. Fever and nuchal rigidity gradually subsided, and spinal fluid approached normal. He was discharged on January 18, 1947, and has remained relatively well since that time."

When exhibited to an audience at the hospital on July 26, 1948, the patient was obviously alert, happy, and very energetic. No neurological effects were seen. The child has grown to a fine young man, who has taken his proper place in human society.

Similar reports soon began to arrive from all over the world. These were often preceded by appeals for streptomycin to save children's lives. In some cases the appeals came too late or there was not enough of the drug available to comply with all requests on time. Many letters began to arrive either from the parents of the patients or from the sufferers themselves. Here is a typical letter received in 1947:

"It was of great interest to me to read that you are the discoverer of streptomycin. To me this is a magical name because the doctors tell me that this is the drug that was responsible for saving my baby's life. I gave birth to a baby girl on March 12th of this year. When she was three weeks and three days old, she took very ill and was rushed to the hospital. Various tests proved that she had spinal meningitis and the doctors held out very little hope for her—they had never had a case of an infant so young contracting this disease. They administered streptomycin and it was this drug which pulled the baby through. Prior to the discovery of this drug, the doctors told me, this disease was fatal in infants."

The drop in tuberculosis mortality among children is well il-

TABLE 5.
TUBERCULOSIS MORTALITY AMONG CHILDREN UNDER
15 YEARS OF AGE IN NEW YORK CITY, 1930-1959

Year	Deaths	Rate[a]
1930	400	24
1935	214	13
1940	110	7.4
1945	92	5.9
1950	78	4.7
1953	47	2.7
1959	14	0.7

[a]Per 100,000 population
SOURCE OF DATA: Drolet, personal communication.

lustrated in table 5. The death rate for the group under twenty-five years of age had declined from 27 per 100,000 population in 1936 to 14 in 1946; the drop became precipitous with the advent of chemotherapy, declining to less than one per 100,000 in 1956 and still less in 1961. Although the tuberculosis death-rate declined for all age groups, the greatest reduction took place in the younger age groups. With the advent of chemotherapy there has also been a significant decline in the number of tuberculosis deaths occurring in the age group of sixty-five years and over. In general, the annual death rate from tuberculosis as a whole has been reduced in the western world from 500 per 100,000 population at the beginning of the century to about 5 or 10 at the present writing (less than 5 in the United States).

The total decline in tuberculosis mortality from all forms is further illustrated in tables 6 and 7 and in figure 19. It has been greater in the decade of 1947-1957 than in earlier years. Unfortunately, the newly reported cases have not declined as rapidly as the deaths. The rapid decline in tuberculosis mortality is an indication of progress being made in the control of tuberculosis, but the high level of new cases reported suggests that the prob-

TABLE 6.
TUBERCULOSIS MORBIDITY AND MORTALITY IN SELECTED COUNTRIES

Country	New cases		Death rate[a]			
	1957	1960	1937-39	1947-49	1958	1960
Canada	7,662	6,345	56.2	37.1	6.0	4.6
Chile	....	....	253.0	209.9	51.7	53.0
United States	66,437	55,494	50.0	29.9	7.1	5.9
England and Wales	32,669	23,605	63.8	49.4	9.9	7.5
Denmark	1,303	....	39.8	24.4	4.6	4.2
France	43,894	....	118.1	73.5	24.3	22.1
Netherlands	7,490	5,688	44.8	29.9	4.3	2.8
Switzerland	5,811	4,841	83.1	53.7	15.3	....
Spain	....	....	123.5	118.5	27.1	....
Australia	4,035	4,057	39.2	27.6	5.5	4.8
Japan	520,899	488,318	208.9	178.0	39.4	34.1

[a]Average annual rate per 100,000 of population.
SOURCE OF DATA: World Health Organization (1962b).

TABLE 7.

TUBERCULOSIS DEATH RATES BY BROAD AGE GROUPS, UNITED STATES, 1936-1961

Annual mortality per 100,000 population.

Age, years	1936	1941	1946[a]	1951	1956[b]	1961
Under 25	27.1	20.4	14.1	5.0	0.9	0.4
25-44	75.1	55.3	41.4	19.6	6.0	3.8
45-64	80.8	68.4	60.1	37.5	16.6	11.3
65 and over	93.1	85.9	70.5	55.0	34.5	34.7
All groups	55.9	44.5	36.4	20.1	8.4	5.9[c]

[a]Streptomycin discovery announced in 1944.
[b]Isoniazid introduced in 1952.
[c]Figures for 1962 gave an annual mortality rate of 4.9 per 100,000 population.
SOURCE OF DATA: *Tuberculosis Chart Series* (1958, 1961c).

lem has not been fully solved as yet. Attention must, of course, be called to the fact that, in the interpretation of the newly reported cases, these data are affected not merely by the continued incidence of tuberculosis or the number of new cases developing each year, but also by changes in reporting practices and the extent of case-finding activities.

PROBLEMS ARISING FROM USE OF STREPTOMYCIN

Streptomycin was not an unmixed blessing. With its continued and extensive use, new problems arose. I drew attention to some of these in receiving the Passano Award at the May, 1947, meeting of the American Medical Association:

"The use of streptomycin in the treatment of tuberculosis presents certain challenging problems to the microbiologist, the chemist and the clinician. Most significant of these are the development by the tubercle organism of resistance to the streptomycin, and the drug's ineffectiveness against certain forms of tuberculosis. In order to overcome these limitations an intensive search must be made for other agents that would be more effec-

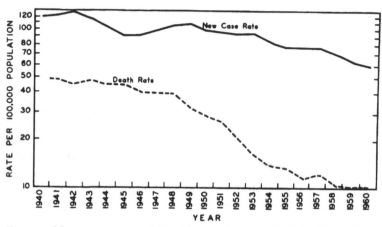

Fig. 19. New-case rate and death rate of tuberculosis in New York City, since 1940. (AFTER LOWELL, 1960.)

tive alone or combined with streptomycin . . . A turning point has now been reached in the chemotherapy of tuberculosis. Although streptomycin may not be the final answer in the treatment of this scourge of mankind—and I hope that it is not—it has opened a new path, a path of antibiotic approach to chemotherapy, an approach sought since the discovery of the bacterial nature of the disease; the control of tuberculosis may finally materialize and thus advance man one step further in his battle against disease and epidemics."

Some ten years after the discovery of the drug, both the progress and the problems of those earlier years were described by Hinshaw (1954) as follows:

"Desperate patients, their relatives, and physicians from many countries throughout the world sought to obtain streptomycin when none was to be had or when it could be put to better use. A committee of the National Research Council under the able leadership of Dr. Chester Keefer bore the brunt of many of these urgent and sometimes hysterical appeals . . . Our clinical advice was sought and gladly given through hundreds of letters, telegrams, cablegrams, and long distance telephone calls during a period of many months . . . Patients were urged to follow the advice of their personal physicians and reminded of the value of conventional treatment methods. Many forms of political and economic pressure were applied to secure the life-saving drug, and a lively international traffic in contraband streptomycin is believed to have developed."

One of the greatest cooperative experiments ever planned in the treatment of any disease was undertaken in 1946 by the Veterans Administration in Washington. The semiannual Streptomycin Conferences, comprising an ever larger number of participating hospitals, contributed, through their exhaustive study of the various methods of treatment of all phases of tuberculosis, to the chemotherapy of the disease. The National Tuberculosis Association and numerous other organizations and hospitals throughout the world became engaged immediately in similar projects, which were both fundamental in nature and of great practical significance in contributing to better understanding of the problems that had now arisen.

P. A. Bunn, assistant chief of the tuberculosis division of the Veterans Administration, in a study of the use of streptomycin carried out in seven veterans' hospitals, reported in January, 1947, on the results of treatment of 91 cases of pulmonary tuberculosis. Twenty per cent showed unexpected or unusual improvement, fifty per cent improved, and only thirty per cent remained unchanged following streptomycin treatment for three months, as shown by x-ray examinations. Fifty patients with miliary or meningeal tuberculosis, or both, were treated: 54 per cent remained alive as a result of that treatment. Complete healing was obtained in practically all of 60 draining sinuses in 12 patients.

Hope Verging on Enthusiasm

Streptomycin was made available to the public at large in 1947. Up to that year, the treatment of the disease depended largely upon bed rest and collapse therapy. The case fatality rates had shown little improvement over the past twenty years. A new era was begun with the introduction of streptomycin, later supplemented by para-aminosalicylic acid (PAS) and isoniazid (INH). Drolet and Lowell were able to state in 1955 that during the preceding seven years, the most rapid decline in tuberculosis mortality ever seen had taken place.

The results obtained in the chemotherapy of tuberculosis through the use of streptomycin, pointing to the final conquest of the disease, were also greeted warmly all over the world by eminent scientists and clinicians. Professor A. Gratia of Liége, Belgium, wrote in 1948:

"Mais il est permis déjà actuellement d'enregistrer cette notion de fait que, pour la première fois, on a trouvé quelque chose qui agit incontestablement, et *in vivo,* sur le bacille tuberculeux, et dans les cas les plus graves. Pour la première fois, une brèche a été créé dans la citadelle de la tuberculose. Cela, c'est une chose vraiment nouvelle et encourageante. Cette notion n'est peut-être pas définitive, mais elle prouve qu'on est dans la bonne voie, sans que l'on réussisse nécessairement du premier coup."

The Italian pediatrician Cesare Cocchi stated in 1948:

"In the first period, December 1946–July 1947, the technique of small intraspinal and intramuscular doses of streptomycin was used together with sulfone and vitamin A. Fifty-five cases of tubercular meningitis have received treatment; twenty-two, definitely cured without subsequent effects, have returned to their normal occupations and have stopped all treatment. Repeated checks have shown that they are not only clinically cured, but also from the point of view of the spinal fluid, their general condition is excellent. Two are on the road to recovery, one still showing partial block; the other has been clinically cured for some months, but the spinal fluid still shows signs of inflammation . . . In miliary tuberculosis of the lungs, the treatment with small doses . . . continued for three to four months, and in some cases even for many months, always cures the patient, and the patient is permanently and definitely cured."

G. Fanconi and W. Löffler of Switzerland wrote in 1948:

"Zum erstenmal in der Geschichte der Tuberkulose ist dem

Artz im Streptomycin ein Mittel in die Hand gegeben, in die Entwicklung dieser Krankheit grundlegend einzugreifen und ihren Verlauf richtunggebend zu beeinflussen . . . Es ist vorauszusehen und zu hoffen, dass unter der Streptomycintherapie das klinische Antlitz der Tuberkulose sich wesentlich ändern und dass es der heranwachsenden Arztegeneration weniger düster erscheinen wird als noch der heutigen."

M. Barber, who inspected medical services in the Soviet Union, wrote in 1951:

"Of the many departments we visited in this hospital (a large children's hospital in Moscow) the most interesting was a 60-bed unit for the streptomycin treatment of tuberculous meningitis. This unit was started in 1946 and it takes cases from all over the Union. The results are said to have improved each year. Two factors were regarded as of major importance—early diagnosis and intensive treatment including injection at the main site of infection. At present an average treatment consists of intracisternal injections daily for about 200 days. By this means it is claimed that the complete recovery-rate has now reached 70%. I understood that over 90% of the cases were bacteriologically confirmed. We talked to many of the children who were recovering and two little girls recited poems to us that they had made up while they were ill."

The efforts of the clinical investigators in this country continued unabated. In 1947, McDermott and others, of the Cornell University-New York Hospital Medical Center, reported that "The administration of streptomycin to 17 patients with bacteriologically proved meningeal, miliary or other types of generalized hematogenous tuberculosis, was followed, in every instance, by a striking alteration in the course of the infection. Six of these individuals (meningitis, two; miliary, two; and other forms of generalized hematogenous disease, two) have attained complete remissions which have been maintained for five to twelve months after the completion of therapy. A seventh patient is apparently recovering one year after an acute miliary infection . . . On the basis of the observations which have been presented, it appears that streptomycin exerts a striking effect upon the course of generalized hematogenous and meningeal tuberculosis . . . In the present study, evidence of therapeutic activity is afforded by: (1) the uniformity with which the administration of streptomycin was accompanied by marked clinical and roentgenologic improvement: (2) the disappearance of tubercle bacilli from those discharges in which they had been easily demonstrable before therapy; and (3) an impressive degree of healing of the lesions in the lung, as revealed by histologic examination . . . Spontaneous recovery in miliary tuberculosis and in tuberculosis meningitis is distinctly unusual." They added cautiously: "The incidence of apparent recovery reported by Hinshaw and Feldman and encountered in this study is unprecedented and is presumably attributable to the action of the streptomycin."

W. S. Tillett wrote, in a highly optimistic mood in 1948:

"It has been an exciting era that has seen a drop in the mortality . . . Since bacterial infections of one kind or another make up such a considerable proportion of the organic ills of man, it is readily understandable that the practical uses of chemo- and antibiotic therapy should be explored to the greatest degree and that published reports of their successes and limitations should fill medical journals in the most conspicuous fashion . . . In viewing the antibiotic age, perhaps greater interest, even if of a more speculative nature, centers around whether or not the present day satisfaction with the success of antibacterial therapy can be transformed into a permanent optimism for generations to come. For an elaboration of this problem it is useful to consider the situation from the point of view of the bacteria themselves. These microscopic forms of viable protoplasm must be viewing with an inordinate degree of alarm the profound threat to their survival that has arisen. The situation to them is of the same magnitude as the age of the atomic bomb is to the human species."

H. McLeod Riggins, in his Introduction to the Riggins and Hinshaw volume on *Streptomycin and Dihydrostreptomycin in Tuberculosis* (1949), stated: "Tuberculosis has been and remains the most destructive and devastating of infectious diseases.

Throughout medical history its challenge has defied all efforts of drug therapy . . . By the fall of 1946, within the short time of a little less than three years after its isolation, streptomycin had . . . successfully withstood the most severe tests in both experimental and clinical tuberculosis . . . For the first time a common objective—the successful treatment and control of tuberculosis—brought together, in a single cohesive group, representative investigators from all parts of the country, supported by . . . different types of organizations."

P. A. Bunn reported in 1949 that "Therapy of tuberculous meningitis with streptomycin has resulted in the recovery of a significant number of individuals and has prolonged the life of many others. Any reduction of mortality in this otherwise fatal infection of the brain and its coverings represents an achievement never before accomplished in the history of tuberculosis."

J. Lorber (1954) summarized the results of streptomycin treatment of 549 consecutive cases of tuberculous meningitis, admitted between 1947 and 1950. The treatment consisted in most cases of prolonged, combined, intramuscular and intrathecal administration of streptomycin.

"The overall two-year survival of 101 patients admitted in 1947 was 31.6 per cent, and of 448 patients admitted in 1948-1950 was 49.3 per cent. Later deaths were very rare. Of 100 patients admitted in 1948-1950 in the early stage of the disease, 74 per cent survived; of 198 intermediate cases, 54 per cent survived; and of 125 advanced cases, 24.8 per cent survived. In 1948-1950, 52.2 per cent of 287 patients without miliary tuberculosis survived, and 44.1 per cent of 161 patients who had miliary tuberculosis and meningitis survived . . . The 245 survivors were followed for two and one-half to six years. In all, 64 per cent had no sequelae whatever. Seventy-five per cent of those admitted in the early stage, 60 per cent of those in the intermediate stage and 40 per cent of those in the advanced stage had no sequelae. In 14.7 per cent of all survivors, deafness was the only residual damage, and two-thirds of these were totally deaf. It is presumed that this complication was usually a consequence

of the streptomycin treatment and was not caused by the meningitis."

Finally, Hinshaw (1954), speaking on the tenth anniversary
of the discovery of streptomycin, said: "Since the discovery of
streptomycin, tuberculosis death rates in the United States have
been reduced to approximately one-third of their former toll. No
physician who cares for patients with tuberculosis will doubt that
modern therapy deserves a large share of the credit for the survival of these many thousands of persons. We, who by chance
happened to be involved in this development, are humbly grateful
for a privilege which comes to few men." Further information
on the early uses of streptomycin is given by Riggins and Hinshaw (1949) and by MacLeod (1956).

Caution Verging on Skepticism

These fantastic developments in the treatment of a disease
that did not previously lend itself to any form of therapy except
rest, climate, and diet were bound to arouse a certain degree of
skepticism. The suspicion of any claims for new therapeutic
agents can be envisioned from the following quotation culled
from the comprehensive treatise of A. R. Rich (1951):

"Whether tuberculosis will cease to propagate itself, as some
now believe, or whether, after reaching a certain level compatible
with the status of the living standards that may prevail, the curve
of the mortality rate will flatten out, and remain more or less
constant, fluctuating with temporary variations in the living
standards, remains for the future to disclose. One thing, however, we should keep clearly before us. In spite of the justifiable
gratification with which we may regard the great decrease in
the *total* mortality from this disease, it is nevertheless very important not to lose sight of the disturbing fact that tuberculosis
is still by far the most common cause of death in that valuable
age period between fifteen and forty-five. In the United States,
in 1946, tuberculosis was responsible for 16 per cent of all deaths

in persons between 15 and 24 years of age, and for 15 per cent of all deaths among individuals between 25 and 40 years of age. It was either the first or the second cause of death between the ages of 15 and 45 in more than half of the states of the union. *The disease that still kills more than twice as many individuals as any other single cause of death during this particularly productive and enjoyable period of the life span can hardly be jubilantly regarded as being 'nearly conquered.'* "

Although published in 1951, the above conclusions seem to have been based upon the 1946 results, or based upon the very first and earliest cures resulting from streptomycin therapy.

John Barnwell, in presenting the Trudeau medal of the National Tuberculosis Association to J. Burns Amberson in 1952 reminded his audience of the difficulties in making an accurate evaluation of any tuberculosis treatment by clinical methods:

"The clinical testing of any therapeutic agent for human tuberculosis and the formulation of a judgment of the results give pause to the physician who has dealt at length with the disease. A recollection of the multitude of remedies, often vaunted by very conscientious men, that have had their little day and then passed into obscurity sometimes sorely tries one's faith in the value of clinical observation. First of all is the notorious tendency of the disease to fluctuate naturally through all the phases of progression and retrogression, sometimes because of treatment; sometimes for the lack of it; often despite it. It requires the utmost watchfulness to avoid the mistake of ascribing causal relationships to events which are merely coincidental, since tuberculosis is more deceptive in this respect than are many other diseases. The difficulty is multiplied if one relies solely for his judgment on the use of a therapeutic agent as a part of the routine procedure in an institution . . . Tuberculosis in animals never duplicates exactly all the features of the disease in man, and, therefore, there is not only a place but a real necessity for clinical testing."

In accepting the Trudeau medal for 1953, J. J. Waring was still concerned with the erratic behavior of tuberculosis. He quoted E. L. Trudeau as saying that "It takes a bit of the Divine Fire to get well of tuberculosis." He further quoted the British doctor Kingston Fowler, who said: "No fool ever gets well of tuberculosis." Waring still felt that the previous general pervading spirit in relation to tuberculosis should be "to cure sometimes, to comfort always." David Cooper, however, in his presidential address before the meeting of the Trudeau Society in 1953 was more sanguine: "For the first time in the history of mankind the possibility of the eradication of tuberculosis is a reality." David T. Smith, speaking at the same meeting, also took a broader view and emphasized the changing the goal from "the control of tuberculosis to the *elimination* of the disease from the United States."

This enthusiasm aroused among the clinicians was such that even Waring himself had to admit that whereas, twenty years before, the annual meeting of the National Tuberculosis Association was "a very quiet, pleasant affair . . . the present meeting not only presented a wonderful program but also gave evidence of the promised land of eradication of tuberculosis."

CONCLUSIONS

The control of tuberculosis has thus become a reality. In spite of all the fears and suspicions, fully justified, at the beginning of chemotherapy, this disease responded, as did other infectious diseases produced by bacteria, to the action of antibiotics and other chemotherapeutic agents. The battle was not fully won, however, by streptomycin itself, since it still had certain limitations as an ideal chemotherapeutic agent. But it was the first truly effective drug. Others were bound to come in time to supplement it or even to replace it.

The treatment and approaching conquest of tuberculosis by man has thus undergone five distinct stages: (1) the health-resort era, (2) the bed-rest era, (3) the collapse-therapy era, (4) the resection era, and now—(5) the chemotherapy era. Each era has overlapped the others. "The conquest of tuberculosis will be recorded in history as one of Man's greatest triumphs against suffering and death from disease . . . Tuberculosis has become a medical disease. Formerly the physician was only an interested bystander and comforter, watching and attempting to aid the forces of nature by improving environmental influences and prescribing symptomatic remedies—the era of the tuberculosis health resorts. This was followed by a period of great dependence upon bed rest, a regimen prescribed like a medicine. A refinement and added feature of local lung rest called collapse therapy was in vogue for many years but disappeared very rapidly. Meanwhile, antibacterial drug therapy—the first truly logical treatment—came to the fore and has retained its prominence." (Hinshaw and Garland, 1963.)

« 8 »

Personal Experiences of Effectiveness of Streptomycin on Tuberculous Meningitis and Other Forms of Tuberculosis

World War II was over. Both the victors and the vanquished began to gather the cultural and economic shreds that were left and to attempt a recovery. To their great joy, they discovered that during the war years the humble saprophytic microbes had contributed a means, quite unexpected, for the rapid recovery of people from infectious diseases. The great potentialities of the antibiotics and particularly of streptomycin to attack disease-producing bacteria, including the tuberculosis organism, had in the meantime become universally recognized. Only few, notably in the United States and in Great Britain, had followed the progress made during the war years. Now all countries of the world were anxious to find out and to take advantage of the new discoveries, the new agents for combating infections and epidemics.

I began to receive invitations in ever increasing numbers from all parts of the world. These came uninterruptedly for a period of nearly ten years, from the end of the war until about 1955. I could accept only a limited number of such invitations to deliver lectures, to visit hospitals, to talk to physicians and patients, to hold conferences, and to advise industrial organizations. This made it necessary for me to undertake many trips abroad, including two around the world, and to visit some fifteen different countries and numerous states in this country.

I thus had an ideal opportunity to observe the effects of antibiotics upon infectious diseases in general and of streptomycin, alone or supplemented with synthetic compounds, on tuberculosis in particular. The clinical cases that impressed me particularly were those afflicted with tuberculous meningitis, and often with miliary tuberculosis as well. This was particularly true of the children. They were indeed pathetic and helpless. Now a ray of hope came into their young lives.

How can I describe the impressions left upon me by the first sight of a child, no matter in what country and in what position in life, who had been saved from certain death by the use of a drug in the discovery of which I had played but a humble part? For the first time in human history, these children, afflicted with tuberculous meningitis and miliary tuberculosis, which once would have meant certain death, now had a chance to survive. They were being restored to life by a drug produced by a soil-inhabiting microbe and discovered in a small agricultural laboratory.

At first there was still not enough streptomycin available for wide distribution, and clinicians still lacked experience in its administration. For these reasons, some of the early cases were brought back to life for only a brief period of time, or were left with serious aftereffects, such as permanent deafness or lack of balance. But more often their lives were saved, and many became, in time, normal human beings.

Who would have dreamed, before 1944, that this could happen? And in such a brief period of time! Professor Giuseppe Caronia, an eminent clinician of the University of Rome, told me that in a study of all the cases of tuberculous meningitis that could be traced throughout human history only fifty spontaneous recoveries had been recorded up to the advent of streptomycin. Even those cases, moreover, may not have been properly diagnosed, and the particular ailments may actually have had other, less severe, causes. When treatment is started early enough, 85 to 90 per cent of all tuberculous meningitis patients are now being saved thanks to streptomycin and the other antibiotic and synthetic compounds since discovered.

TABLE 8.
Tuberculosis Mortality among Children under
15 Years of Age, Various Countries, 1950 and 1959

Country	Deaths		Death rate[a]		Decline
	1950	1959	1950	1959	(per cent)
Canada	466	51	11.45	0.88	92
United States	1531	240	3.78	0.44	88
England and Wales	909	72	9.45	0.69	93
France	1771	208	19.40	1.80	91

[a]Per 100,000 of child population.
Source of Data: Drolet and Lowell (1962).

Who could describe the anxiety of the parents of those children, or their happiness upon seeing their dear ones return to life from certain death? In those early days, soon after the discovery of streptomycin, when little was known of its purity and of its possible side effects upon various human organs, many a father, who had been forewarned of the possible undesirable reactions, notably deafness, said to me, "I would rather have a deaf child than a dead child." (See table 8).

Sir Charles Harington, of the Medical Research Council in Great Britain, told me in 1947 of some of his experiences when the first lots of streptomycin began to arrive at his laboratory. It was highly essential to test its effectiveness in experimental animals, in order to establish the toxicity and dosage of the drug, before attempting its use in human beings. Upon learning that he was in possession of this lifesaving drug, many parents of children afflicted with the dreadful disease of tuberculous meningitis came to him to beg him not to waste the wonder drug on animals, but to use their children for the experiments. The investigator had to possess very strong will-power to withstand such appeals.

I would like to tell the reader about a few of these early experiences so that he may have some concept of the profound impressions they left on me.

1. THE STORY OF FATHER ST. PAUL

One of the first tuberculosis patients being treated with streptomycin who was brought to my attention was a Jesuit priest, a delightful person, who was professor of medieval history at a great university. His tuberculosis was already in an advanced stage; one of his kidneys had already been removed, and the other had become infected. He was given only about six weeks to live when he was transferred to the Mayo Clinic. When he learned of the advent of streptomycin and its effectiveness in experimental tuberculosis, he begged to be used in the very first experiments on the clinical evaluation of the drug. The only streptomycin at that time available was made up of small amounts produced in culture flasks in my own laboratory and at Merck and Company. The process of its chemical purification was still not fully worked out.

I received occasional reports from the Mayo Clinic concerning the condition of Father St. Paul. Although he apparently developed vestibular disturbance (sense of balance) owing to the effect of the unpurified form of the drug, his health improved to such an extent that he was soon asking to be allowed to return to his home in order to resume his teaching. In May, 1946, or nearly a year and a half after his treatment was begun, I visited Rochester, Minnesota. On arrival at the hospital, I was taken immediately to visit him. He had just returned from Mass in company with a younger priest. They were sitting at a table in a very jovial mood. When I was introduced to him, his face lighted up. He took both my hands in his, and exclaimed: "This is the happiest moment of my life. I have been praying to the

Almighty that I should live long enough to meet you and express to you my personal gratitude, which should be added to that of many others whose lives are being saved by your marvelous discovery."

When we were seated, he began to unfold to me his life history: how he had become sick; how he had spent his time in various hospitals; how he finally had come to this clinic, apparently for the sole purpose of being allowed to pass here the remaining few days of his life; how he had first heard of the discovery of streptomycin; how he had begged the doctors to use him as a kind of guinea pig; how he was treated, first intermittently, since the supplies of the drug were still arriving at the clinic only at infrequent intervals; how he had begun to feel a gradual improvement in health with each new course of treatment. He then led me triumphantly to a large closet. Since he was a highly intelligent patient, he was allowed to keep his own record of the temperature changes, blood condition, bacteriological analyses, and other pertinent data. As he unrolled the long paper, on which his record was carefully written down, he pointed out the periods when his treatment with streptomycin was started and the subsequent effects; when it was stopped, there were the resulting changes in his condition; then it was started again, followed by other changes. "Now," he said, "I feel that I should be allowed to return to my daily tasks. I am ready to resume my normal way of life, thanks to you and to your streptomycin." As he pressed my hands, his eyes were radiant.

To say that I was elated would hardly express my own feelings. Rather, I felt very humble to think that the lowly microbes, the actinomycetes, to the search and study of which I had devoted so many years, were able to produce such marvelous substances that could do so much good for the suffering members of the human race. Even if only this one human being were allowed to live, to make it possible for him to continue to impart his wisdom to others a while longer! Perhaps the time had come when younger lives, still on the threshold of their existence, could be saved, so that they, too, might lead normal lives.

As I expressed my hopes to this fine, humanitarian scholar, he blessed me and wished me greater success in my efforts. The younger priest watched my departure, his eyes, too, shining and his face lighted with hope for a healthier and perhaps a better world.

Unfortunately, Father St. Paul's treatment was only partly successful, since he died a year later.

2. MOSCOW IMPRESSIONS

In the summer of 1946 I accepted an invitation from the Russian Academy of Sciences to deliver in Moscow a series of lectures on the subject of antibiotics. As we were making ready to leave for Europe by airplane, a journey involving at that time, because of prevailing postwar conditions, many unforeseen difficulties,[1] I began to receive one telegram after another to the effect that certain very important persons in the Soviet Union had been afflicted with tuberculous meningitis and other forms of tuberculosis; I was requested to bring some streptomycin with me.

At that time the supply of the drug was still so limited that the special committee appointed to supervise its distribution, both here and abroad, released a quantity only upon the recommendation of medical experts for the treatment of each individual case in order to keep careful records of the results following its use. The drug could thus be obtained only through official channels. In some way, however, small supplies of streptomycin of an unknown degree of purity found their way into the black market, and these supplies were being sold in Europe at fabulous prices. Nobody, of course, could guarantee the nature and efficacy of such material.

The morning after our arrival in Moscow, my wife and I were invited to visit the Children's Hospital, to see Ninotchka, an

[1]These were described at length in my book, *My Life with the Microbes,* published by Simon and Schuster, New York, 1954.

eight-year-old child, daughter of an eminent mathematician of the university. We were told by the doctors that when the child was admitted to the hospital, twelve weeks previously, she was given only about three weeks to live. A small amount of streptomycin was obtained through black market channels; she was treated in the best way known to the local physicians at that time. The result was an immediate and remarkable improvement. Ninotchka soon seemed to be on the way to recovery. Unfortunately, the supply of the drug gave out, and the treatments became intermittent and irregular. The variable dosage used, the uncertainty of proper administration, and the known impurity of the drug had affected her hearing.

Upon our arrival at the hospital, we were led to a well-lighted room with a dozen or more beds occupied by children suffering from various diseases. We found Ninotchka sitting up in her bed and having a late morning meal. The group surrounding the bed included, besides my wife and myself, some ten or twelve doctors, as well as the father of the child. As soon as she saw her father, a broad smile spread over her flushed face. The doctor in charge began to explain to us the method of treatment and the various stages of improvement of the patient. Some of us wanted to question her as to her condition. Because of her apparent loss of hearing, the questions were written out on a small blackboard that she kept on her bed. She answered them all in a clear, almost singsong voice. As he watched his only child and listened to her recital, the father beamed. Grasping my hand, he thanked me profusely for saving her life. I expressed my most profound sympathy to him and my sincere regret on her loss of hearing.

Meanwhile, the doctors surrounding the bed were listening with much astonishment to the report of the physician in charge. Only a great painter could have done justice to the scene before us. For the first time in his entire experience, every doctor present was watching the return to life of a patient condemned to certain death by the dreadful affliction. The chief physician then turned to me and said: "As far back as our records can show, there have been brought to this hospital from all over the country, annually, more than two hundred children suffering from tuberculous meningitis. But never has a single child left this hospital alive. This is the first time that I am witnessing such a recovery. It is a historic moment for me, for the hospital, and for the medical profession as a whole. We shall always remember this day and this occasion!" I felt very humble indeed.

Everyone asked anxiously, "How can we obtain more streptomycin to complete the saving of this child and help other children as well?" Representatives of the Soviet Union had been trying, with only moderate success, to purchase all the streptomycin they could find, in whatever amounts available. Unfortunately, the quantities that could be spared, even in the United States, were still very limited. A year or more was to pass before sufficient streptomycin would become available for export.

I did not see Ninotchka again, since the doors of the Soviet Union were soon closed to the outside world. I heard through different channels, however, that she was gradually improving, and that she was leading a normal life. When we again visited Moscow, years later, we were told: "She is in good health, developing normally, attending the University, and successfully meeting her school requirements, but she remains deaf."

During our visit to Moscow in 1946 we met the great navigator and scientist, Professor Otto J. Schmidt. This was the man who a few years before had made a voyage to the North Pole, a voyage that has since become famous. He spent the winter on an ice floe, from which he and his party were rescued by American fliers and brought to the United States. The navigator was received at the White House by President Franklin D. Roosevelt, an event fully described by the navigator himself in a widely read book. Evidently, during that winter he contracted pulmonary tuberculosis, which was now endangering his life.

We were invited to Schmidt's apartment for lunch, where we found ourselves in the midst of a group of eminent scientists and physicians. The navigator was a tall, heavy-set man with prominent facial features and a full beard. He and his wife greeted us heartily in typical Russian fashion. We were requested at once to take places at a table laden with food and drinks. The ensuing conversation dealt largely with tuberculosis and streptomycin. The host kept leaving his seat every few minutes to expectorate into a special cloth which he kept in a spacious pocket. Although the effects of streptomycin on pulmonary tuberculosis were still little known at that time, the physicians felt certain that if they could obtain a supply of the drug, our host would be greatly benefited by treatment with it. The life of the famous navigator was thus placed in my lap, so to speak. I could only promise that upon my return home I would try my best to help obtain a sufficient amount of streptomycin for his case.

As we said good-bye, the large blue eyes of the naviagtor looked into mine with an expression of mingled hope, confidence, and appreciation. What could I tell him? Encourage him? I could hardly do that, since I always tried to refrain from making promises which I might not be able to keep. I could only say that every effort was being made to produce the drug in larger amounts, and express the hope that some of it would reach him in time to be of help to him. For years, interrupted communication with the Soviet Union closed the door to any further information. But on a subsequent visit to Moscow (May, 1957) I learned that the navigator had improved on treatment with streptomycin, had almost regained his normal health, and had lived for another eight or ten years to make some highly important contributions in the field of earth sciences.

3. THE CHILDREN OF PARIS

The year 1946 was a historic one. In spite of many scoffers, the fact that chemotherapy of tuberculosis was possible came to be gradually recognized. Throughout the world, ever increasing numbers of cases, representing various forms of the disease, were being treated with streptomycin with varying degrees of success. Numerous reports began to appear in the United States and in other countries on the effectiveness of streptomycin. Many of these reports were enthusiastic, some were cautious, and a few were still skeptical. This skepticism was pronounced in certain European institutions. Although I was welcomed in the various hospitals and laboratories which I visited in 1946 and 1947, I could sense considerable caution. When an eminent physician asked me frankly: "Do you claim that streptomycin is a cure for tuberculosis?" I tried to assure him that this was quite beyond my province. I could tell him all about the antibiotic, its origin, chemical nature, activity upon various bacteria, including the tuberculosis organism, in the test tube and in the experimental animals; the rest was left to the pathologists and clinicians, like my friends Feldman and Hinshaw. (See table 9).

TABLE 9.
TUBERCULOSIS MORTALITY OF PERSONS 15
YEARS OF AGE AND OVER, 1950 AND 1959

Country	Deaths		Death rate[a]		Decline
	1950	1959	1950	1959	(per cent)
Canada	3,210	908	32.9	7.8	76
United States	32,428	11,189	29.2	9.2	68
England and Wales	15,060	3,783	43.8	10.8	75
France	22,515	10,279	68.8	30.9	56

[a]Per 100,000 of adult population.
SOURCE OF DATA: Drolet and Lowell (1962).

There were exceptions, however. One of these I encountered in Paris in 1947. There, an invitation awaited me to visit the famous Hôpital Salpetrière, where Jean Fouquet had treated several children for tuberculous meningitis.[2] He took me through the wards and introduced me to a number of patients. I talked with them about their ailments and the kind of treatments they were receiving. Streptomycin was beginning to be used in small

[2]These cases were later described by Fouquet and his associates (1949). The first cases of streptomycin-treated tuberculous meningitis in children in the United States were described by E. M. Lincoln in 1947 and later in 1961. A more detailed study was made in France by Debré and Brissaud (1953).

amounts. When I was ready to leave the hospital, Fouquet said to me with a smile, "Please do not hurry, we have a little surprise for you." Attendants then brought in two children, a boy five years old and a girl eight years old. They were dressed in colorful provincial costumes and each carried a large bouquet of flowers.

"These children are Michel and Janet," said the physician. "They were brought to this hospital six months ago, from distant regions of France. They were almost in a state of coma upon arrival. We began at once to treat them with streptomycin. As you see, they have both made a complete recovery. All the tests for residual tuberculosis appear to be negative. We were going to send them home last week, but we heard that you were coming to Paris and we have kept them here so that you could see them yourself. Are they not pretty?"

With these words, he presented the children to me. The little girl curtsied. Both children kissed my hand, handed me the flowers, and I patted their lovely, curly heads. As I leaned over them, I felt like crying. These two children had, surely, been saved by streptomycin! For how long, was still a question. Some such afflicted children would no doubt be completely cured and would lead a normal life; others in time might face a recurrence, or suffer permanent injury to their hearing or their sense of equilibrium. It was still too early to say definitely what the future would hold for them, since disturbing reports were coming in concerning recurrences and various vestibular and auditory reactions. The medical profession was still very much on guard. The clinicians were still afraid to speak of "cures." Not enough was yet known of the mode of administration of the drug or of the amounts to be used. Much still remained to be learned. But one thing was certain; many of the children, and grown-ups as well, had survived. Here were two wonderful witnesses. One could only hope for the best.

These thoughts were interrupted by Fouquet's request that I step into the courtyard of the hospital where one of the interns was anxious to take a photograph. This was the first photograph in which I posed with children saved by streptomycin. On leaving the hospital, with the flowers in my hand, I looked back to see the two little ones waving good-bye. This was enough compensation for all the sleepless nights and the endless days spent in the study of the lowly microbes of the soil.

Two years later (in 1949) Fouquet published a summary of the results of streptomycin treatment of a number of children in France. In sending me a copy of this report, he wrote:

"Grâce à vous, la méningite tuberculeuse n'est plus la maladie inexorable qu'elle était il y a trois ans à peine, et il faut bien dire que les résultats obtenus ont largement depassé les espérances que l'on exprimait au début de cette nouvelle thérapeutique."[3]

4. A VISIT TO THE ZURICH STATE HOSPITAL

On our second postwar visit to Europe, in 1947, we came to Switzerland to rest after a series of scientific conferences in England, Denmark, France, and elsewhere. Here in Switzerland among the quiet mountains and lakes, away from hospitals, scientific gatherings, and industrial organizations, one could attempt to summarize the numerous observations just made of the efficacy of streptomycin in the control of disease in general and of tuberculosis in particular.

We stopped in Zurich to accept an invitation by Professor W. Löffler, director of the tuberculosis division of the local state hospital. Some thirty tuberculous patients were in the clinic at that time and were all being treated with streptomycin. They varied greatly in age and in the nature and condition of the disease. I conversed with each of them, either in French or in German. I was particularly interested in their physical and mental states, and especially in their response to the treatment with strepto-

[3]"Thanks to you, tuberculous meningitis is no longer the horrible disease that it was barely three years ago; one must say that the results already obtained have far surpassed any hope that one could have had at the beginning of this new form of therapy."

mycin, watching for any possible undesirable reactions. All the patients were cheerful and grateful. None complained.

Some were being injected with the drug, one into the spinal fluid at the base of the brain, for which an opening had to be drilled. This patient was a young man about twenty. As he was being treated, he emitted painful sounds, not so much an expression of deep suffering as a kind of complaint. Not being familiar with the clinical behavior of this type of patient, I asked whether these exclamations signified any special form of pain. The nurse giving the injection smiled at my innocent question. The patient, she said, was being encouraged to utter such sounds, in order to guide her in the speed with which she was allowing the lifesaving drug to enter his brain. One cheerful little woman, of an indeterminate middle age, kept smiling and telling me how much better she felt now than when she first entered the hospital a few weeks previously.

When we returned to his office, Löffler summarized the various cases for me. Then he added, with a smile: "You wanted to know my opinion about streptomycin. All I can tell you is that half of the patients to whom you have just spoken would not be alive today were it not for your antibiotic." He dwelt particularly upon the favorable effects of streptomycin on the morale of the patients, as well as on their health. They stopped coughing, their temperature returned to normal, and they had a general feeling of well-being. He finally concluded his remarks by saying: "Whether streptomycin is the final answer to the treatment of tuberculosis or not, and we all hope, as I am sure you do, that it is not, nevertheless, we are all convinced that effective chemotherapy of tuberculosis is now possible. This gives us, both the clinicians and the patients, courage to carry on. I can truthfully say that streptomycin has opened the door to the therapy of tuberculosis, the ancient foe of the human race, and has revolutionized the treatment of this disease."

These were profound words, full of meaning. Climaxing other impressions gained from other types of patients and in other countries, what I saw here stimulated my own morale. The life-

saving properties of streptomycin were a reality after all! It is
to be recalled that this was in 1947, when the only detailed studies
of the effect of streptomycin in experimental and clinical tuberculosis were still being made largely in the United States. The
British were still skeptical. The French had had little experience. The northern countries were still to be heard from. Russia
had little to go by. I came away from that hospital more cheerful than I had been for some time.

5. THE CHILDREN OF SEVILLE

Nearly three years passed between our 1947 journey to Europe
and the next voyage. These were years of great consolidation of
the results obtained with streptomycin. Knowledge of the effect
of this drug in the treatment of numerous infectious diseases,
especially tuberculosis, was rapidly accumulating. Hundreds of
reports were being published. Many books were being written.
During this three-year period, I had to attend numerous conferences throughout the United States, deliver various addresses,
write many articles for scientific journals, edit a comprehensive
volume on the subject of streptomycin, direct an ever increasing
number of research workers in my own laboratory, and reply to
numerous letters from all parts of the world about antibiotics in
general and streptomycin in particular. Meanwhile, I continued
the search, with the help of several research assistants and graduate students, for new antibiotics that would have the capacity
to attack other diseases and possibly take over where streptomycin left off. A new antibiotic, *neomycin,* was isolated in 1948;
we were now spending much effort in establishing its place in
the armamentarium of the medical profession, especially its possible use in tuberculosis to supplement or replace streptomycin.

Early in 1950, another journey to Europe was in order. The
immediate purpose was to attend a meeting of a committee
called together for the coördination of the production and utiliza-
tion of antibiotics throughout the world. This meeting, in Geneva, Switzerland, was to be held at the request of the World

Health Organization. Afterward we traveled widely in western Europe. Everywhere, especially in Spain, numerous letters from tuberculosis sufferers awaited me. Parents whose children had benefited from the use of streptomycin paid us visits. Often they brought the children with them. They expressed joy and thankfulness and frequently shed tears of appreciation.

As I entered the lecture hall of the University of Madrid, where I was to deliver an address, an elderly lady, dressed in black, handed me a package. It contained a book of poetry, with an extensive inscription. This read, in translation, as follows:

"Undreamt of are the mysteries of Divine Providence! I have heard your voice over the National Radio of Spain. As I have a sad heart and a soul full of pain, your words moved me very much, and I am very grateful to you.

"I had six sons. I lost my three oldest ones, victims of pulmonary tuberculosis, and the three younger ones victims of tuberculous meningitis. To think that if you had discovered streptomycin a few years before, I could still have the great joy of having my sons with me. In thinking about them, I cried very much.

"As this book is dedicated to the women of the world and as it has the virtue of having an effect upon the heart, I would like to greet your wife, giving to both of you the picture of a mother who is sad because of human pain, and who will dedicate every day a thought and a message of sincere and everlasting gratitude to you."

I was told later that this lady was a famous poetess in Spain.

Wherever we went, especially in Andalusia and in the Basque country, we were met by enthusiastic crowds, dancers in the street, cannon salutes, and public officials doing us honor.

The feelings of the man in the street can best be illustrated by the following story. During our brief stay in Seville, I entered a small haberdashery. A Spanish scientist who had accompanied me to act as interpreter mentioned my name to the storekeeper. The man immediately rushed from behind the counter and embraced me, shouting in highly emotional language, and with tears running down his face, how his only son had been saved by streptomycin from certain death by tuberculous meningitis. I naturally offered my best wishes, shook his hand, and tried to pay him for my purchase.

He refused to accept the money from me, saying, "I cannot accept any payment from the saviour of my son's life. It is a great and unexpected honor for me to serve you." I then said jestingly, "Suppose I were to ask for your whole store, would you be willing to give it to me?" His immediate reply was: "If Jesus Christ had walked into this store and asked me for it, would I refuse it to him?" There was so much humility, such profound religious feeling in these words, that I could only feel humble.

My Spanish friend and I insisted that the storekeeper accept some compensation for the purchase, since he did not seem to be a wealthy individual. He finally agreed to take a certain amount, but not more than the actual cost, without any profit for himself. As we were about to leave the store, it occurred to me that I needed a box to pack some dresses presented to my wife by a group of gypsy dancers the night before. When I asked the man whether he could spare a cardboard carton, he promised to send one to the hotel. When we returned there, we found a simple leather bag awaiting us.

Several children came with their parents, who tried to express, in a language unfamiliar to me, their gratefulness because of the saving of their loved ones. These people, especially some of the little girls and boys, made a profound impression. Similar incidents occurred in Granada, in Bilbao, and wherever we went in Spain. There was everywhere deep feeling, almost religious in nature. What could I do but bow my head in silence?

6. THE CHILDREN OF VERONA AND ROME

From Spain we proceeded to Italy. Here we found (this was still in 1950) that considerable progress had already been made in establishing the clinical effectiveness of streptomycin. Several important clinics in Rome, Florence, Verona, and elsewhere were busily engaged in studying the effect of this drug, now being supplemented with PAS or with certain vitamins. Everywhere we met assurances of the blow that had been dealt to tuberculosis and especially to tuberculous meningitis.

The Florentine pediatrician, Dr. Cesare Cocchi, had used streptomycin in his clinic as early as December, 1946. At first, owing to improper or insufficient dosage, the results were not very encouraging. Gradually, however, he came to the conclusion that "at last we have at our disposal a powerful and effective remedy that has already changed and will further change the fate of millions of patients." Before long he was able to say that "tuberculous meningitis can now be completely cured." In 1950, he began to speak of "the total elimination of tuberculosis," saying, "We have at our disposal remedies which in time to come will make tuberculosis only a sad relic of the past."

A group of physicians from Verona visited us in Florence and invited us to come to their city to see what had been accomplished in the use of streptomycin for the treatment of tuberculosis. On arrival there, we were greeted by a group of children, ranging in age from six to eighteen years, who had been saved by streptomycin from certain death from various forms of tuberculosis. A poem written by them for the occasion was read to us by a small, curly-headed boy. A bronze statuette, a copy of the famous statue of Con Grande, the Duke of Verona, was presented. It bore the following inscription: "SW, benefattore dell umanità gli amma lati dell ospedale fracostoro di Verona, con riconoszenza. Verone —26 Maggio 1950."[4]

Another demonstration of a similar nature took place a few days later on our visit to the famous Forlanini Institute in Rome, where the clinician Dr. Omodei Zorini reported to us on the results of his studies on the chemotherapy of tuberculosis with streptomycin. We found here a well-organized system of statistical recordings of the morbidity and mortality rates from tuberculosis in various countries. A large hospital offered beds to sufferers from various forms of the disease. We were shown the

[4]"SW, benefactor of humanity . . ."

marvelous results obtained by the use of streptomycin and the extensive experimental work that was going on in combining this drug with other antituberculous agents. As we left the hospital, a large group of patients who had benefited from treatment with streptomycin came out of their rooms to give us a most touching ovation. Other results of a highly encouraging nature were presented at the Children's clinic near Ostia by the Roman pediatrician Professor G. Caronia (fig. 20).

FIG. 20. A group of tuberculous Italian children treated with streptomycin at the Children's Hospital in Ostia, with Professor and Mrs. Caronia (on the outside) and Dr. and Mrs. Waksman, 1950.

7. BRITISH IMPRESSIONS

The British were skeptical at first. The possible reactions from the use of this drug were kept in mind and even overemphasized.

Because of the shortage of available streptomycin and the numerous demands for it, broadcasts were actually being made in 1946 over the British Broadcasting Company by the Minister of Public Health, Aneurin Bevan, warning against the indiscriminate use of streptomycin and emphasizing its potential toxicity. The *British Medical Journal* wrote in 1946 that "there seems to be a very real risk that, even if the infection is controlled, the patient will usually be left mentally deficient, deaf, blind, or otherwise a hopeless invalid." Soon, however, more favorable reports began to appear.

Professor John Lorber, of the University of Sheffield, reported in December, 1949, the results of treatment with streptomycin of ten children suffering from tuberculous meningitis: "All the children were alive and well at the end of July 1949, after a minimum observation period of one year from the beginning of treatment. All the children are in good physical health and are normal mentally. There is no evidence that the illness has caused any difference in their general behaviour. . . . The only neurological after-effect is deafness in one case."

8. A VISIT TO THE LAENNEC HOSPITAL IN PARIS

The year 1950 was memorable for visits to numerous hospitals where we observed scenes similar to or even more striking than those previously recorded. This was true of several countries, and especially of France. During the three years since we were there last, the importance of streptomycin in the treatment of tuberculosis had become firmly established. There was now great excitement concerning the efficacy of this drug in controlling various forms of tuberculosis. The leading physicians of Paris had now come to recognize fully the merits of streptomycin as a chemotherapeutic agent. Two French companies had already undertaken to manufacture this drug on a large scale, thus making it readily available for treatment of the average patient.

Two occasions have impressed themselves on my mind particularly. The first was a visit to the headquarters of the International

Union against Tuberculosis. Here I addressed a large gathering of eminent scientists and clinicians. Among them was C. Guérin. He and A. Calmette had devoted their lives to the development of a vaccine against tuberculosis. This vaccine, which has had world-wide use, is called BCG (Bacillus Calmette-Guérin). At the end of the conference, the aging scientist came up to greet me in words that can be freely translated as follows: "Now that we have finally succeeded in obtaining a highly promising drug for the therapy of tuberculosis, my own humble contribution directed primarily toward preventing this disease begins to take a second place."

Paris is a city famous for its great scientific developments. Here it was that the great French phthisiologists, notably Laënnec and Villemin, whom I have discussed at length in a previous chapter, did their epoch-making work on tuberculosis. Here it was that the great master of microbiology, Louis Pasteur, made his historical studies on rabies and other infectious diseases. Here the eminent microbiologists Emile Roux, Emile Duclaux, Elie Metchnikoff, and other disciples and followers of Pasteur uncovered the causes of various infectious diseases of men and of animals. Here, too, it was that numerous other chemists, biologists, and clinicians carried out their work, beginning with Lavoisier's classical studies on fermentation and continuing into Claude Bernard's investigations in human physiology. I considered it, therefore, a great honor to be received in this city and to be made a Commander of the Legion of Honor and Member of the French Academy of Sciences.

Etienne Bernard, the head of the tuberculosis division of the famous Laënnec Hospital, extended an invitation to visit the clinic and examine the results of their work. Later I was requested to address, at that hospital, an assembly of eminent representatives of French medicine. In the wards of the famous hospital, patients who were being treated with streptomycin seized my hands, pressed them, and uttered exclamations of thankfulness and appreciation. Doctors had frequently mentioned the uplift of the morale of the patients as a result of my visits to tuberculosis sanatoria and special hospitals. But the logical Frenchman is usually not known for such sentimental and passionate outbursts as I witnessed here. The patients were apparently prepared for my coming, for a small delegation of the hospitalized patients presented me with several beautiful volumes on modern French art and with bouquets of flowers for my wife. As we left the wards, waving good-bye to the patients, thunderous applause followed us.

At the conclusion of my address at this hospital, the French clinician and writer Georges Duhamel expressed the appreciation of the scientific world in general and of this audience in particular for the discovery of streptomycin. He emphasized, rather unnecessarily I thought, my European origin. My reply was that science is international in spirit, that an idea may originate in one country, be developed in another, come to fruition in a third, and finally find application throughout the world, including the very country in which it originated.

9. ATHENS

The primary purpose of a visit to Athens, Greece, in 1952, was to receive an honorary doctor's degree from its university. We were greeted, on arrival, by a number of clinicians and professors, who had much to tell about their experiences with streptomycin in the treatment of various forms of tuberculosis. Among the hospitals and sanatoria we visited was the children's clinic of the university, where Professor K. Choremis, in charge of the tuberculosis work, showed us through the wards. The efficacy of the drug was by then well established. The best modes of administration for the various forms and stages of the disease were well understood. After leaving the clinic, we were invited to rest in the courtyard of the university, under the spreading branches of an old olive tree. Tea and other refreshments were served. Then Choremis exclaimed: "Now comes the real surprise that we have in store for you!" As he said this, we saw approaching a middle-aged, simply dressed couple, leading by the hand a pretty little girl, seven or eight years old, with curly blond hair and dressed in light blue. The child came up to us and curtsied. The parents kissed our hands, bending deeply over them.

"This is Happiness," said Choremis. "She is the first child in Greece to have been saved by streptomycin from tuberculous meningitis. The parents were so happy to have their only child come back to life that they have changed her name to 'Happiness.'" I placed my hand on the child's head and was happy myself. Without a knowledge of Greek, I could only smile and shake hands with the parents, who rewarded me with bows and words that I could only sense were expressions of gratitude.

The next day, I delivered an address at the university before a large gathering of students, physicians, and teachers. As I looked over the surrounding hills, especially toward the majestic Acropolis and the ancient Agora, I thought of the great Greek philosophers and dramatists, of Socrates and Plato, of Aristotle and Empedocles, of Sophocles and Euripides, who had trod these very grounds nearly twenty-four centuries ago. It was they who laid down the philosophy of life and the principles of human relations to their environments that were to influence human thought for thousands of years to come. I thought particularly of Hippocrates, who, together with his followers and assistants, laid the basis for our understanding of tuberculosis and its treatment. The thought occurred to me: "No knowledge of microbes existed then. These organisms have since come a long way. And now comes a simple searcher for soil microbes to bring to the inhabitants of this ancient city a message gained from a study of these lowly microbes, inhabiting the very earth trod by these great thinkers who lived right here, in Athens. These scholars hardly suspected that the earth under their feet harbored the solution for the ancient 'phthisis.'" The thought then came of other ancient scholars, in another culture and in another land, not too distant, who proclaimed just about that time in the Book of Ecclesiasticus that "The Lord hath created medicines out of the earth; and he that is wise will not abhor them."

That evening our Greek friends took us to the theater to witness a performance of *Prometheus Bound* on the very spot where this tragedy of Aeschylus was first produced. The audience, upon learning of our presence, stood up and gave us a rousing ovation. It was indeed breath-taking. I could not help comparing the act of Prometheus that unleashed the power of fire for man's use, one of the greatest blessings conferred upon man, and the more recent work of all those searchers for the unknown who have helped to unleash the potential properties of the microbes of the earth, thus giving man other tools for combating his mortal enemies, those that cause infectious diseases.

10. STOCKHOLM

Another highly significant event took place later that year, in December, during the week of the Nobel prize celebrations in Stockholm. The National Tuberculosis Association of Sweden arranged a dinner in our honor. At the table, on my right, sat Countess Bernadotte, who represented the Queen of Sweden, honorary president of the Association, and, on my left, Professor Nanna Svartz, an eminent clinical investigator and the only woman member of the Nobel prize committee. It was a Sunday evening affair, three days before the ceremonies of the formal award of the prize.

After the dinner, several addresses were delivered by some of the leaders in the field of tuberculosis in Sweden. Finally my turn came to make the response. I told them of some of my personal experiences in the use of streptomycin in the treatment of various diseases, especially different forms of tuberculosis. Among these experiences, I said, I would like to mention one that had happened only that morning, right there in the city of Stockholm: "The manager of the hotel telephoned our room to say that a gentleman and a little girl were downstairs anxious to see me. Since they spoke no other language than Swedish, with which we were not familiar, the manager offered to come up with them and serve as their interpreter. In a few moments they arrived. The man was middle-aged, apparently a workingman, dressed in his Sunday best. He was leading a little girl seven or eight years old by the hand. The child carried a bouquet of red carnations. With a curtsy, she handed me the flowers. The father of the child, whose name was Hellström, took my hand and began to talk quietly. The manager translated his words, which were an expression of thankfulness that the life of his child had been saved from tuberculous meningitis five years earlier. I could only appreciate his coming to see me and in bringing the child with him."

As I finished telling this story, one of the clinicians said: "Yes, indeed, I remember that case well. I treated the child just over five years ago. She was apparently the first tuberculous meningitis sufferer in Sweden whose life was saved by streptomycin."

This was not the end of the story. A newspaper reporter soon learned of what had happened, and early the next morning, appeared in our room wanting further details about the father and the child. Since I did not know who they were, other than their name, I referred the reporter to the physician who treated the child. To my great surprise, immediately after the Nobel ceremonies in the Concert Hall, the reporter rushed up to me with the little Hellström girl. She was now dressed in her holiday best and carried in her hand a fresh bunch of five red carnations, one for each year of her newly gained life. As she handed me the flowers, a corps of photographers suddenly appeared to take pictures. The next day the newspapers carried the story, which practically overshadowed that of the Nobel prize event. We were invited to a dinner given that evening by the King of Sweden for the prize winners and their wives at the Royal Palace. After the dinner, the party broke up into small groups, each carrying on a lively conversation about the events of the past few days. The royal couple mingled with the guests. As the King approached our little group, he congratulated me and said, "You must be very proud of having received, for your work on streptomycin in tuberculosis, the greatest prize that a scientist can hope for." My reply was: "I am very grateful indeed. But I also received that day an even greater prize, one from a child whose life was saved by streptomycin." I told him then the story of the five carnations and added, "What greater prize can there be than that of saving a human life?" The King agreed with me.

11. JAPAN

With the Nobel ceremonies completed, we flew to Tokyo, where I was to deliver addresses in several Japanese cities in connection with the celebration of an important scientific event in that country. We left Stockholm early in the morning of December 14, and fifty-two hours later landed at the Tokyo airport. Here we were met by a crowd of people, made up of scientists, community groups, and interested individuals. There were literally hundreds of photographers and reporters, as well as a number of kimono-dressed girls, to greet us.

During the twenty-two days of our stay in Japan, we heard much about tuberculosis and streptomycin. Several industrial organizations were now producing the drug in large quantities. The annual death rate from tuberculosis in Japan was said to have dropped to less than one third of the 1941 rate, and the decline was being ascribed largely to the use of streptomycin. The story of the treatment of a Hiroshima child suffering from tuberculous meningitis, for whom Norman Cousins had pleaded several years previously, was well known.[5] We were told of a number of other experiences, and I met many of the people involved.

One morning the desk clerk at the Imperial Hotel in Tokyo telephoned to our room to say that a Japanese gentleman and a small boy were there to see me. Since we had been receiving numerous visitors, I was inclined to refuse this request. The clerk insisted, however, that the man would not leave the hotel until he was permitted to see me. The best that could be done was to go down to the lobby to find out what it was all about. As I approached the desk, a neatly dressed middle-aged man, accom-

[5]"Hiroshima—Four Years Later," *Saturday Review*, 32(38):8-10, 30-31, 1949.

panied by a boy of ten or eleven years, came up and began to bow in the fashion of the country, repeating words that I could not understand. The clerk came to my aid by translating. It seemed that the visitor's only child, the boy who was with him, had been saved from certain death from tuberculous meningitis by streptomycin. He had read in the newspapers about my coming to Tokyo, and wanted to express his son's best wishes. I shook hands with both of them and wished them well. Still, they would not leave. The father kept repeating that he was a tailor and would like to make me a suit and present it personally as a token of his appreciation. I was greatly embarrassed. I really did not need any additional clothing and would not have been able to find space for another garment in our overcrowded bags. It took much effort and considerable conversation to convince the father that I was grateful enough for his coming to see me, without the garment.

Other boys and girls came to express their feelings of thankfulness. Some wanted me to visit them in their country homes, and I accepted a few of these invitations. Everywhere, little presents were handed to me as tokens of appreciation.

12. THE CASE OF DR. P

The final story may be devoted to the experience of Dr. P. He was a young surgeon who came from Great Britain to the United States early in 1945 to carry out special work in medicine at one of the leading medical schools in the country. Soon after his arrival, he became afflicted with miliary tuberculosis, one of the worst forms of the disease. He was given only a few weeks to live. Since there was no hope for him, he was sent to a sanatorium in the mountains, where he could live and die in peace.

Just about that time, a pharmaceutical company was planning to undertake the manufacture of streptomycin. The officials of the company were anxious to determine whether the vestibular reactions that frequently followed the administration of the drug then (1945) were produced by the active substance or by some accompanying impurity. It was decided to prepare a kilogram of crystalline streptomycin and place it in the hands of a group of eminent clinicians for the careful evaluation of any possible side reactions of the drug. At the prevailing price levels, this material was worth more than $25,000. The chief of the clinic selected for this study was familiar with the case of Dr. P, since it was he who had sent P to the mountain sanatorium. He immediately decided to invite P's coöperation in this particular experiment.

P agreed and returned to the hospital. As soon as the streptomycin became available, he was selected as one of the first human "guinea pigs" in this important experiment. He was fully aware of the risk involved and the potential consequences. The experiment was highly successful. P regained his health and returned to normal life. Unfortunately, his vestibular nerve was affected and he lost his sense of balance control. Nevertheless, he was eventually able to overcome this handicap to a large extent by compensatory use of his other senses and when last heard from he was leading the life of a successful practicing physician.

These, then, are a selected group of personal experiences. During the years 1946-1952 most of the experimental and basic studies on streptomycin were completed. By 1950 it was established beyond question that streptomycin had delivered a deadly blow to the "white plague" in all its forms. Most of the undesirable aftereffects that appeared during its earlier use were being largely corrected by changing the mode of administration, by reducing the concentration, by the supplementary addition of PAS, and later (in 1952) by the introduction of isoniazid (INH).

In 1952 when I was introduced to the Emperor of Japan by the American Ambassador, Robert Murphy, the Emperor said to me (through an interpreter): "In 1941, the number of deaths from tuberculosis in Japan was three-hundred for every hundred thousand of our population. Last year (1951-1952), the mortality dropped to ninety. We like to ascribe this to the effect of streptomycin. But what I would like to know is what the chances are that the rate may be reduced still further, perhaps to zero."

My answer was that streptomycin was the first drug to establish the potentialities of chemotherapy in tuberculosis. We were all hoping that it would not be the last drug. Others would, no doubt, come to supplement it, and perhaps even to replace it as more effective. At the time we were discussing this problem, experiments were being carried out in the United States on the use of isoniazid, alone and in combination with streptomycin, in the treatment of tuberculosis. I was, therefore, most happy to read, six years later, that the annual mortality from tuberculosis in Japan had dropped from 90 to 45 per 100,000. This rate has been reduced still further in 1960, as shown in table 6.

«9»

New Antibiotics and Antituberculous Synthetic Compounds Supplement Streptomycin

The chemotherapy of tuberculosis, or its treatment by means of chemical agents, thereby resulting in the arrest of the disease process, came into its own in the years 1945-1947. This manner of treating the ancient affliction of man is to be differentiated sharply from the previously prevailing "cures," as well as from those which involved sanatorium care, with its bodily rest, hygienic measures, and pure physical or mechanical treatments, including collapse. Chemotherapy is also to be differentiated from various nonspecific forms of therapy, such as the use of drugs for the sedation of cough or for the stimulating of the appetite. Streptomycin demonstrated that chemotherapy of tuberculosis is possible. It proved to be a specific agent highly effective against this disease. (Previously only a slight effect has been obtained by the use of sulfone drugs.) Streptomycin was soon supplemented by other antibiotics and certain synthetic compounds that have found an important place in the treatment of tuberculosis.

The discovery of the action of streptomycin upon various forms of human and animal tuberculosis aroused considerable excitement in both the medical and lay worlds. Coöperative studies were immediately initiated on a scale never experienced in the history of medicine. One of these studies was begun in the United States by the hospitals of the Veterans Administration and those of the Army and Navy. The results thus obtained were at once reported at conferences called together for the purpose. They served as a basis for further investigations and further evaluation of this and other drugs. The first "Streptomycin Conference" was held in Chicago on December 12-14, 1946. This was followed by three conferences in 1947 and two each in 1948 and 1949. When new antituberculous drugs were introduced, their comparison with streptomycin was at once undertaken. The subject of the annual conferences was then changed to "Chemotherapy of Tuberculosis"; later they were designated "Research Conferences on Pulmonary Diseases." They continued to be held in different cities, with ever greater numbers of hospitals represented.

At these conferences, reports were presented dealing with various regimens used in the treatment of different forms of tuberculosis, at first the use of streptomycin alone and later its use in combination with other antibiotics and synthetic compounds. Supplementary to these semiannual and annual conferences, the Veterans Administration prepared frequent reports dealing with the results in the various coöperating hospitals. New recommendations and new procedures, based upon these reports, were thus made available through immediate wide distribution. Other agencies in this country, as well as in several European countries, and the International Union against Tuberculosis, with headquarters in Paris, also undertook systematic studies on the chemotherapy of tuberculosis and submitted their reports and recommendations. The British Medical Research Council has been very influential and has designed and executed some of the best scientific clinical studies to demonstrate the special values of streptomycin.

Industrial organizations were not slow in taking up the cues presented by the coöperating hospitals and other clinical bodies. Many companies began to manufacture streptomycin on a large scale and to initiate broad research programs of their own, both on experimental animals and in clinics. This resulted in the isolation of numerous new antibiotics and synthetic compounds, several of which proved to be effective in the treatment of tuberculosis. Some of these soon became of practical importance in supplementing streptomycin and in some cases in replacing it where the latter failed because of resistance or toxic reactions.

In a relatively short time, covering roughly the decade and a half between 1945 and 1960, the chemotherapy of tuberculosis was definitely established. Its remarkable development was the result of: (*a*) the rapid progress in the use of antibiotics in the treatment of numerous infectious diseases caused by various bacteria and other microbes; (*b*) the enthusiasm thus aroused and the prevailing expectation that drugs would soon be forthcoming that could be used in the treatment of tuberculosis; (*c*) the extensive and careful coöperative efforts in which the Veterans Administration, the Public Health Service, the medical section of the National Tuberculosis Association, the Tuberculosis Chemotherapy Trials Committee of the British Medical Research Council, the Trials Committee of the Swedish National Association against Tuberculosis, and numerous others participated.

In analyzing this rapid progress in the treatment of a disease which only twenty years before was believed to be incurable, and which was thought to be completely resistant to true chemotherapy, Long (1958) gave credit, first of all, to the advent of synthetic sulfones, which were considered to be the first break in the long, previously fruitless studies of the therapy of tuberculosis. The discovery of 4,4'diamino-diphenyl-sulfone was the logical sequel to the sulfonamides. The work of Feldman and Hinshaw with the gluco-sulfone (promin) was the first to show that experimental tuberculosis of the human type in guinea pigs could be arrested (and cured) by chemotherapy. (Further references may be found in Feldman, 1954, and in Hinshaw and Garland, 1963). These results opened the way for the still more effective synthetic compounds that came to be known as promin, promizole, and diasone. This led Domagk and his associates to suggest the use of thiosemicarbazone drugs in the treatment of tuberculosis.

These sulfones were readily water soluble, heat-stable, and apparently nontoxic. On oral administration, they were tolerated fairly well by man and animals. Attempts were immediately made to use these compounds for tuberculosis in man. At first the results seemed to be encouraging; striking improvement was obtained in recent lesions of the exudative type; in many cases the sputum became negative for tubercle bacilli. Unfortunately, the toxicity of these compounds soon became a disturbing factor; secondary anemia and agranulocytosis resulting from their use induced the clinicians to proceed rather slowly with their further use. It may be of interest to mention that because of their effectiveness against acid-fast bacilli it occurred to Dr. Faget of the Carville leprosarium that they should be tried in human leprosy —an experiment which has led to a revolution in the treatment of this great plague of man.

According to Long, drugs that are effective in the treatment of tuberculosis should possess the following qualifications: toxicity for tubercle bacilli, at a range much below that required in the treatment of man; capacity to penetrate all the infectious lesions of the disease; ease of administration, whether by mouth or by parenteral methods; a persisting bacteriostatic effect, in spite of mutation of the infecting organism; and a low capacity to stimulate allergic drug sensitivity. The prevailing concept of the mode of action of such drugs is that they interfere with the normal metabolism of the invading bacterial cells, a concept for which little proof is yet at hand.

Although the sulfones and other early synthetics aroused hopes of something better to come, the first really powerful antituberculous drug (streptomycin) came from an unexpected quarter, the saprophytic microbes, notably the actinomycetes which inhabit our soil and composts in great abundance. Although streptomycin had its limitations, it pointed the way toward the final solution of the tuberculosis problem. This is brought out in table 8, which shows the steep drop in the mortality of tuberculous children since 1950.

The remarkable progress made in the reduction in the tuberculosis death rate among children was partly, if not largely, owing to the reduction of the transmission of the disease from person to person. The death rate for the age group under twenty-five years declined from 27 in 1936 to less than 1 per 100,000 population in 1956. The relative decline in the death rate for the older age group, sixty-five years and over, has been said to be largely owing to the increasing population in this age group; there has been a remarkable drop in the mortality from tuberculosis in this group as well since the advent of drug therapy.

The courses of morbidity and mortality from tuberculosis in the United States are shown in figures 21-23. Although the rapid drop in the death rate began in the period 1945-1948, the years when streptomycin was introduced, and became accelerated in 1952, when isoniazid (INH) was introduced, morbidity rates have receded more slowly.

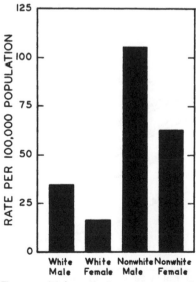

Fig. 21. Tuberculosis case rates and number of cases in the United States by race and sex in 1959. (AFTER *Tuberculosis Chart Series*, 1961c.)

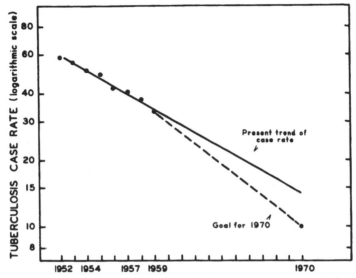

FIG. 22. The decline in the tuberculosis case rate. (AFTER *Tuberculosis Chart Series*, 1961c.)

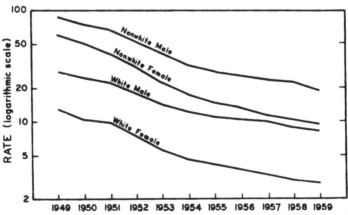

FIG. 23. The trend in the tuberculosis death rate by race and sex. (AFTER *Tuberculosis Chart Series*, 1961c.)

Death rates from tuberculosis in other countries tell a different story. The changing rate in Poland is illustrated in figures 24 and 25. It is to be noted that the rise in mortality in 1915 was coincident with World War I and the Russian Revolution. Another rise took place between 1940 and 1945 as a result of World War II. The rate at which new cases were reported dropped more slowly than the mortality rates. Mortality and morbidity changes in Brazil are shown in figure 26. The less favorable results in Brazil following the introduction of antituberculosis drugs may be ascribed to their indiscriminate use. An examination of the resistance of 2,182 cultures from patients, carried out in the Central Laboratory of Tuberculosis in Rio de Janeiro between Janu-

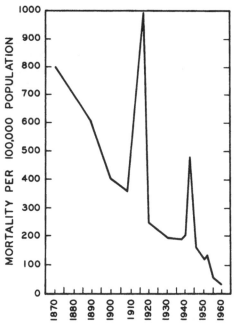

FIG. 24. Mortality from tuberculosis in Poland from 1870 to 1960, per 100,000 of population. (AFTER HORNUNG, 1962.)

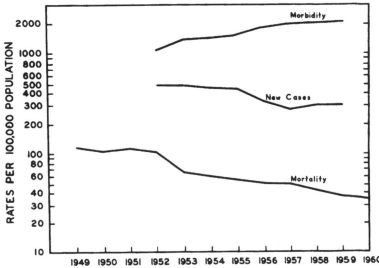

FIG. 25. Morbidity incident (new cases) of tuberculosis in Poland between 1949 and 1960, per 100,000 of population. (AFTER HORNUNG, 1962.)

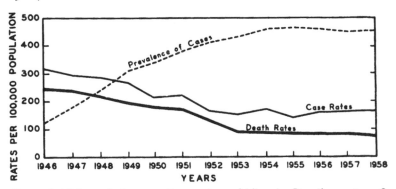

FIG. 26. Tuberculosis mortality and morbidity in Brazil, 1946-1958. (AFTER MAGARAO *et al.*, 1961.)

ary, 1958, and September, 1959, showed that a high proportion was resistant to the major drugs. Two thirds were resistant to two or all of the three standard drugs and 90.6 per cent were resistant to at least one of them; only 9.4 per cent were sensitive to the three (Magarão et al., 1961). The conclusion was reached that "the fact that less than 25 per cent of the cases in Rio de Janeiro are being cured is mostly due to microbial resistance, which causes the specialized hospitals to have the best part of their beds occupied by chronic patients who are sputum positive and resistant." This could have been prevented by properly supervised combined chemotherapy.

New Antituberculosis Antibiotics

The chief limitations to the extensive use of streptomycin in the treatment of tuberculosis were: (a) the tuberculosis bacteria were not killed rapidly by the drug—they were prevented from multiplying and after some time their reproductive capacity was greatly reduced; (b) on prolonged treatment with the drug, and after the long-term therapy which is essential for this type of disease, the development of resistance among the surviving bacteria took place; (c) a peculiar specific toxicity of the drug for the eighth cranial nerve led to a loss of hearing or the sense of balance in some patients; (d) the necessity for administration of the drug by injection because it was not well absorbed by the body when taken by mouth. The reduced form of streptomycin, dihydrostreptomycin, was also characterized by a potentially greater ototoxic effect upon the human body.

To meet these objections an intensive search was undertaken for new effective antibiotics and for synthetic compounds that could supplement or replace streptomycin, especially when the organisms became resistant to it or when it became necessary to avoid possible toxic manifestations.

A large number of antibiotics were soon isolated from cultures of actinomycetes, as well as from fungi and bacteria; many of them were found to be active against the tuberculosis organism. On continued experimentation, some of them found practical application in human therapy. It is sufficient to mention neomycin, viomycin, the tetracyclines, cycloserine, streptovaricin, kanamycin, and capreomycin. There were many others.

ANTITUBERCULOUS SYNTHETIC COMPOUNDS

The first important synthetic compound that proved effective against the tuberculosis organism and came to supplement streptomycin in the treatment of the disease was para-aminosalicylic acid, usually shortened to PAS. Its discovery resulted from a purely theoretical study of the effect of a closely related compound, p-aminobenzoic acid, on the metabolism of the tuberculosis organism. On the basis of the results of these studies Lehmann (1946, 1949) was led to test a large number of benzoic and salicylic acids having similar inhibitory effects upon the bacterium. Among the various chemical compounds tested, para-aminosalicylic acid proved to be the most effective, a concentration of 0.15 mg per cent causing a 50 to 75 per cent inhibition of the growth of the organism *in vitro*. The effect of this compound was next tested in experimental animals and was found to be effective, notably in guinea pigs and rats. Of particular importance was the fact that the substance could be given orally as well as intramuscularly or intravenously. Its effectiveness upon pulmonary tuberculosis in man was then investigated. The response was rapid when the drug was given by mouth, in doses of 10 gm. daily, for a period of three months. When given alone or in combination with streptomycin, PAS also gave good results in cases of extrapulmonary and intestinal tuberculosis.

The second, and even more important, effective synthetic compound discovered was isoniazid. This substance was found in 1952 to be a highly potent tuberculostatic drug. Although isoniazid was first synthesized in 1912, it remained of purely academic interest for nearly forty years. Its antibacterial effect upon *Mycobacterium tuberculosis*, both *in vitro* and *in vivo*, was demonstrated in 1951 in the laboratories of three pharmaceutical houses (Farbenfabriken Bayer, Hoffmann-LaRoche, Inc., and E. R. Squibb & Sons), all independently and at about the same time. The discovery of the effectiveness of this important chemical compound was largely a result of a follow-up study of the antituberculous properties of chemical derivatives of thiosemicarbazones. Several investigators have studied the tuberculostatic activities of the hydrazide of isonicotinic acid, which was used as an intermediate in certain of the chemical processes leading to the synthesis of the pyridine analog of amithiozone.

Upon treatment of tuberculosis in man, the high efficacy of isoniazid was immediately demonstrated. The effects of isoniazid were described by Long as follows: "Fever declined and other manifestations of tuberculous toxicity disappeared, patients gained weight, improved in appetite, and experienced a renewed sense of well-being. Cough and expectoration decreased and in the course of time sputum, after months during which it was positive for tubercle bacilli, failed to disclose these organisms. Radiographic clearing required longer time, but it too was unmistakable after a few months, both in cases with widespread tuberculous pneumonia and in more chronic cases with fibrosis and excavation. An unequivocally favorable effect was also evident in the grave forms of tuberculosis, miliary disease, and tuberculous meningitis."

Oral administration of the drug, in doses of 3 to 5 mg. per kilogram of body weight, showed within a few hours plasma and cerebrospinal fluid concentrations 20 to 80 times as great as that required for an inhibitory effect upon the growth of the tuberculosis organism.

It soon became evident, however, that the administration of isoniazid alone also tended to favor development of bacterial resistance, just as in the case of streptomycin. A study was, therefore, undertaken of associated therapy of isoniazid with streptomycin or with PAS. This therapy proved to be highly effective and soon became standard practice. The Veterans Administration, the Public Health Service, and others established the fact, however, that for short-term-treatment results isoniazid alone was as good as any other single drug or drug combination. It can be administered orally and is rapidly absorbed from the alimentary tract; it can also be used by parenteral administration.

Other synthetic compounds effective against tuberculosis were later isolated. They include ethionamide and pyrazinamide, the most efficient secondary drugs. These find application where other antituberculous drugs have failed.

MacLeod stated in 1956 that for patients under thirty years of age 1 gm. of streptomycin daily plus 100 mg. of isoniazid twice daily or 1 gm. of streptomycin daily plus 5 gm. of PAS four times daily were highly effective regimens. To prevent the effect of streptomycin on the eighth nerve and the development of giddiness, it was recommended that older patients should be given 1 gm. of streptomycin twice or thrice weekly plus 100 mg. of isoniazid twice daily plus 5 gm. of PAS twice or four times daily. Under this form of chemotherapy, the tubercle bacilli should be cleared from the sputum within six months even in the presence of cavities, and virtually no resistant organisms should emerge. It was also stated that most patients should have returned to work in twelve to eighteen months. The use of a single chemotherapeutic agent was said to be "bad treatment." "Salvage treatment" was used when resistance had developed to one or more agents; it consisted of a combination of the drug to which the organism was still sensitive, together with viomycin, oxytetracycline, ethionamide, or pyrazinamide. Combined therapy deserves particular emphasis.

THE CONTROL OF TUBERCULOSIS

The introduction of chemotherapy in the treatment of tuberculosis, owing to the discovery of various antituberculous drugs, notably streptomycin and certain other antibiotics, on the one hand, and of PAS, isoniazid, and certain other synthetic compounds, on the other, resulted in a death blow to tuberculosis, the "white plague" of man. With improved sanitary measures, better care, and proper diagnosis, combined with prophylaxis and chemotherapy, mortality from the disease has gradually declined. It was soon recognized that, as a result of these developments, before long we will be able to speak of the complete control of tuberculosis.

These conclusions become evident from an analysis of the reports from various centers dealing with the mortality and morbidity of tuberculosis. Drolet and Lowell wrote in 1955:

"With the tuberculosis death rate in 1954 in the United States down to 10 per 100,000 population, it is difficult to realize the extreme mortality rates which prevailed not so long ago. In 1912 . . . among operatives in the cotton mills of Fall River . . . tuberculosis showed an extraordinary excess, the death rate averaging annually 320 per 100,000. Among girls aged fifteen to twenty—219, young women twenty to twenty-five—304, those twenty-five to thirty—443; and in the age group thirty to thirty-five, it reached the unenviable peak of 473.

"So rapid have been the changes in the prevalence of tuberculosis and in its mortality during the past few years, especially since 1947, that it has become important to establish 'bench marks' and to record carefully what has been happening . . . Improvement in the treatment of tuberculosis through the development of antimicrobial drugs and more successful thoracic surgery has been the basic event in the recent history of the disease. Let us first recall that, while streptomycin was discovered in 1944 and subjected to careful clinical study in 1945 and 1946, it was primarily in 1947 that this life-saving drug became generally available. Second, it was in 1949 that para-aminosalicylic acid (PAS), developed abroad, likewise came into hand here. The latter, while not too effective alone, in combination with streptomycin contributes advantageously by delaying the emergence of resistant strains of tubercle bacilli and thereby lengthening the therapeutic effect of the drugs. Third, it was in 1952 that the general use of isoniazid became possible . . . One may, therefore, speak of the period from 1947 to 1953 as the first seven years of the chemotherapeutic era in the treatment of the most serious infectious disease affecting mankind for centuries. And, the great successes of surgery in tuberculosis during those same recent years have been generally ascribed to the new effective drugs, improved anesthesia, and improved surgical techniques. Prior to 1947, the treatment of tuberculosis had been largely dependent on bed rest and collapse therapy and . . . the case fatality rates had shown but little improvement for more than twenty years."

In New York City, with a population of eight million, the number of deaths from tuberculosis dropped in 1957 to 982 and the number of newly active cases to 6,117. The death rate for tuberculosis has continued its steady downward trend. In his annual report for 1959, Lowell (1960) spoke of "new gains against tuberculosis" in New York City: "New milestones have been reached and there is hope that, with the present rate of progress, in a generation or two we may see tuberculosis as a comparatively minor cause of illness and death. Tuberculosis is no longer one of the ten leading causes of death. This in itself is an achievement of major importance. Whereas only a few years ago the annual mortality was measured in the thousands, today we count the deaths in hundreds." Lowell cited the 779 deaths recorded in 1959, compared with 833 in 1958 and 2,321 in 1950. While the death rate fell during the year by 7 per cent, the number of new cases of tuberculosis reported in the city fell by 10 per cent. This is brought out dramatically in figure 16 and in table 4.

In speaking of tuberculosis mortality among children, in a report boldly designated as "The Last Stage," Drolet and Lowell (1962) noted that "Children have benefited most from the campaign against tuberculosis. A study of mortality records in four different countries reveals a unique record of life-saving during the last decade—greater than has ever occurred in so limited a period. In four countries: Canada, the United States, England and Wales, and France, 4,677 deaths from various forms of tuberculosis were reported in 1950 among children fifteen years of age in a total child population of 63,299,000. Only ten years later, namely, in 1959, the number of such deaths was but 571 while the child population had increased . . . to a total of 82,135,500. In the United States, the deaths from all forms of tuberculosis among children under fifteen years of age numbered 1,531 in 1950 [and] only 240 in 1959, while the child population was rising by 14,500,000 to a total of 55,227,000. The death rate from tuberculosis, 3.78 per hundred thousand in 1950—the lowest on record in the four countries under consideration—was reduced to 0.44 in 1959 or by 88 per cent."

In 1961 Drolet[1] was able to state: "In New York City, the number of deaths fell between 1950 and 1959 from 78 to 14, or by 82 per cent, when the child population was rising by 233,703. In 1930 these deaths numbered 400. In England and Wales these deaths decreased by 92 per cent. In the Scandinavian countries the decline was also strikingly high; 86 per cent in Denmark, and 95 per cent in Sweden; in Norway, not one death."

Drolet and Lowell emphasized particularly the recovery rates from tuberculous meningitis:

"Before the advent of chemotherapy, tuberculous meningitis was practically always fatal. Lincoln reported in 1947 that, of 92 cases in which a diagnosis of primary tuberculosis had been established and in which tuberculous meningitis developed while they were under observation or later follow-up, 66 died in Bellevue Hospital and another 23 succumbed to meningitis in other hospitals. The remaining three children died at home shortly after the diagnosis had been established in the hospital." Debré in 1952 was already able to report a recovery rate of 25 per cent among infants and 62 per cent among older children in a study of 262 patients in France. Lincoln in 1960 reported a 71 per cent recovery rate even when the meningitis was complicated with miliary tuberculosis.

Lincoln emphasized in 1961 that tuberculosis was preëminently a social disease. "It increases where living conditions are poor and homes are overcrowded. Any measures to relieve poverty and its attendant evils of inadequate nutrition and crowding will help in the basic control of this disease. In order to promote the ultimate eradication of tuberculosis the pediatrician must function not only as a physician but also as a public-minded citizen intent on securing for every child in his city or state the right to be protected from a preventable communicable disease."

[1] Personal communication, April 6, 1961.

THE CLOSING OF THE SANATORIA

One of the most significant developments resulting from the introduction of specific drug therapy and the reduction in the necessity for hospital treatment has been the closing of numerous sanatoria and the conversion of many to other uses. Although hospital treatment is still recommended for most patients, particularly when they are just starting their treatment, the stay in the hospital is usually shorter than it was in former years, and at least part of the treatment period can be managed outside the hospital by private physicians and clinics. This has resulted in a gradually diminishing need for sanatoria or special hospitals for the treatment of tuberculosis patients. Whereas previously there were long waiting-lists of patients who applied for admission to such hospitals, now the beds are being emptied at such a rapid rate that some sanatoria and hospitals have been and are being consolidated, converted into general convalescent homes, or closed altogether. Thus, a disease that less than two decades ago was still regarded as the greatest threat to the health and life of man, a threat that hung over the heads of the people like the sword of Damocles has been reduced to the tenth position or even farther back, among the killers of human beings.

As we traveled through Spain in 1954, our hosts kept pointing out beautiful groups of buildings on the outskirts of cities, saying: "These were to be our new tuberculosis sanatoria, of which we Spaniards were justly proud. Now, because of chemotherapy, the buildings will never be finished and the hospitals never completed, unless some other need arises." (See fig. 27.)

Hinshaw reported in 1955: "During these 10 years a 75 per cent reduction in deaths from tuberculosis has been accomplished in the United States. Much of this reduction is attributed to improved therapeutic methods, a fact that can be demonstrated readily in the tuberculosis wards of any hospital. The duration of necessary hospital stay has been reduced by 50% in some institutions, with resulting diminished need for beds—a substantial economic gain."

Hill (1960) was able to say: "As a result of intensive application of these new medications, the Veterans Administration in the period from 1954 to 1959 cut the death rate among its tuberculosis veterans by more than 50 per cent and eliminated the use of 6,000 beds for tuberculosis. Testifying before Congressional committees, V.A. officials estimated that this mass drug evaluation and treatment program had saved American taxpayers a minimum of $100 million."

This diminishing need for tuberculosis sanatoria has been evident all over the world, but particularly in the United States. In 1943 I addressed a group of patients at the Roosevelt Hospital near my home town and my subject was the possibility that in time a "cure" might be found for their disease. This hospital has now become a general hospital because tuberculous patients are too few to justify its maintenance as a sanatorium. The closing of the famous Trudeau Sanatorium, founded in 1880, has made many people more aware than any figures could of the events taking place in the field of tuberculosis therapy.

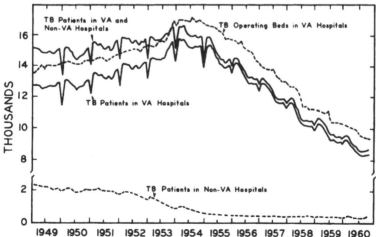

FIG. 27. Tuberculosis operating beds in Veterans Administration hospitals and tuberculosis patients in Veterans Administration and other hospitals, January, 1949–December, 1960. (AFTER *Transactions of the 20th Research Conference in Pulmonary Diseases.*)

The U.S. Public Health Service reported (Tuberculosis Chart Series, 1961) that in 1960 about half of the beds available for tuberculosis patients were used by patients who stayed in the hospital for less than six months, and that three-fourths stayed less than a year. Nevertheless, a significant number of patients stay for long periods because of difficulties in their treatment. This is usually the result of inadequate drug treatment when the disease is first discovered. Many patients do not realize how important it is that they avoid any interruption in treatment over the first year or two.

SUMMARY

During less than twenty years, following the discovery of streptomycin, the mortality from tuberculosis has been dropping at a very rapid rate; this has also been accompanied by a gradual reduction in morbidity. A large number of other antibiotics and several highly active synthetic compounds, subsequently discovered, have come to supplement and, in many cases, to replace streptomycin in the treatment of various forms of tuberculosis, thus contributing in no small measure to the final solution of the tuberculosis problem.

Unfortunately, the morbidity from this disease has not been reduced to so great a degree as the mortality rate. Also, the danger to public health from tuberculosis carriers still lurks around the corner, since many incompletely cured individuals may be walking the streets as possible sources of infection. It is further true that patients often leave hospitals too soon—before they have received the full benefit of hospital care and even before they have become noninfectious. Another problem has been the undue haste of public officials to save money by closing hospitals where a need still exists. These are the remaining problems of a once devastating disease. They will be solved in the years to come.

«10»

The Conquest of Tuberculosis

The final chapter of the battle against tuberculosis appears to be at hand. The progress that has been made in recent years in the control of this disease has been gradual, with rises and falls in the curves of morbidity and mortality. Industrialization and urbanization have contributed markedly toward the rise. Public health measures and general advances in our understanding of the disease and of its infectious nature have contributed toward its fall in mortality. The last stage in this direction had its origin nearly a century ago. At first came the careful diagnosis of tuberculosis, then the isolation of *Mycobacterium tuberculosis,* the causative agent of the disease, the discovery of the X-rays, and surgical measures. Prolonged rest, good food, and fresh air were the only means available for curative purposes, properly designated as "nature's cure." The introduction of the pneumothorax and thoracic surgery have contributed further to the alleviation of the disease. Finally came chemotherapy.

The remarkable progress in the actual attack upon tuberculosis in recent years was a result of two distinct approaches. (1) Measures were at first developed which tended to increase the patient's resistance to infection; this was brought about by means of proper vaccination, and, more recently, by prophylactic use of certain antituberculosis drugs. (2) The discovery of the sulfonamides in 1935 led to the introduction of the sulfones; these had only limited antituberculous action. The discovery of streptomycin in 1944 demonstrated that highly effective chemotherapy of tuberculosis was possible. The limitations of streptomycin, largely because of the emergence of resistant organisms and because of certain toxic side effects, led to the use of combined drug therapy. This was made possible by the introduction in 1949 of PAS (para-aminosalicylic acid), followed by INH (isoniazid) in 1952. Later came other antibiotics and other synthetic compounds.

PRESENT METHODS OF TREATMENT OF TUBERCULOSIS

The drug combinations now used most commonly in the treatment of various forms of tuberculosis are streptomycin and INH, streptomycin and PAS, INH and PAS, or all three drugs together. The choice of the particular combination, the amounts used, and the length of treatment depend entirely upon the nature and stage of the disease.

The tendency has been recently in many quarters to consider INH as the preferred single drug, because it has virtually no toxicity and can be given by mouth. Further, the potential ototoxic side effects of streptomycin require that it be used with caution in elderly people or in the presence of renal impairment. Usually, however, INH is given with streptomycin or another drug. The recent tendency has been to administer daily combinations of INH and PAS. Frequently, streptomycin is also used twice weekly, every other day, or even daily. Many eminent investigators believe that the use of streptomycin in the treatment of tuberculosis will increase.

The period of treatment with drugs, or the period of chemotherapy, is now continued well beyond the time when the "target" point has been reached, namely, sputum conversion, cavity closure, and X-ray stability; it does not represent, however, the "end moment" of chemotherapy. According to various clinical authorities, an adequate course consists of daily combined drug therapy for at least twelve months after reaching target point—i.e., a total of eighteen to twenty-four months of chemotherapy. Indefinite continuation of drug therapy is now recommended for individuals who have advanced disease damage and who are therefore at more risk of relapse. In many instances the extended therapy period is carried out with INH as the only drug. Whether or not this one drug will be sufficient to hold the disease in check permanently is not known.

In developing countries, sputum-positive patients cannot be accommodated in hospitals even at the beginning of their illness. Ambulatory treatment with supervision from a clinic is the only practical course. A major disadvantage is that patients at home cannot be depended on to take their drugs regularly. It is possible in some situations to require clinic attendance at short intervals so that drugs can be given by injection. Many patients have more confidence in injections than in pills. This would seem to be an opportunity for a greater use of streptomycin. Canetti and Rist (private communication) believe that it is better to use massive doses at the beginning of treatment in order to ensure a cure, than to extend treatment for a lifetime. This approach, however, does require hospital care in the beginning and is, therefore, not applicable to the situation in many countries.

Combinations of drugs are definitely bactericidal. Even if a few bacteria survive, they remain in closed foci, and cannot be excreted, and thus represent no danger to other people. If patients are resistant to all the three major drugs or even to one major drug, there are now available other drugs, such as cycloserine, kanamycin, viomycin, pyrazinamide, and ethionamide, that can be used. Chaves and his associates (1961) reported that out of 398 strains of *M. tuberculosis* isolated from ambulatory patients in New York City during 1960, resistance was shown by 75.1 per cent to either streptomycin, INH, or PAS and by

38.2 per cent to all three of the drugs. Hobby (1962*a*) estimated that approximately 75 per cent of tuberculosis patients admitted to municipal hospitals in urban areas within the United States are re-treatment cases; the majority of these patients are infected with tubercle bacteria which are in some degree drug resistant (table 10).

TABLE 10.

PRIMARY DRUG RESISTANCE IN TUBERCULOSIS
PATIENTS UNDER 20 YEARS OF AGE

Year	Total number of strains tested	Per cent of strains partially or totally resistant to		
		Streptomycin	PAS	Isoniazid
1953	84	7.0	3.5	6.0
1954	108	3.0	3.0	4.0
1955	356	1.4	1.4	1.8
1956	285	1.4	3.0	2.8
1957	287	0.7	2.4	2.1
1958	217	3.6	2.3	3.6
1959	209	2.4	1.0	2.9
Total	1,546	2.1	2.1	2.8

SOURCE OF DATA: HOBBY (1962*b*).

The Committee on Therapy of the American Trudeau Society of the National Tuberculosis Association (Barnett *et al.,* 1960) recommended certain dosages for the specific drugs which are now used in the usual cases of tuberculosis, including streptomycin (intramuscularly) and isoniazid (orally). When isoniazid is used in high dosages, it should be accompanied by a minimum of 50 mg. pyridoxine per day to reduce the risk of isoniazid neurotoxicity. Of PAS the sodium form is generally preferred, although its other salts are also effective; when sodium restriction is important, one of the other salts of PAS should be used.

In the case of failures or relapses, five newer drugs are recommended. Cycloserine (orally) in high dosages entails risk of neurotoxicity, which may be modified by the employment of pyridoxine. Pyrazinamide (orally) must be used with caution

because hepatoxicity—liver damage—is an unpredictable complication with this drug. Viomycin (intramuscularly) should be
used with caution in patients with preexisting renal disease. Tetracyclines (orally) and ethionamide (orally) are the other drugs
recommended.

A period of at least eighteen months of uninterrupted chemotherapy, and often a much longer period is required to achieve
maximal benefits. Hospitalization for treatment of failures and relapses is important. When surgical intervention is necessary and
the patient's organisms are resistant to the most effective drugs,
it is often possible to provide a few months' coverage with the
secondary drugs and thus bring him through a critical period.

The effectiveness and relative toxicity of streptomycin versus
dihydrostreptomycin[1] have recently attracted considerable attention. The *Quarterly Progress Report of the Veterans Ad-
ministration-Armed Forces Study on the Chemotherapy of Tuberculosis* stated in April, 1963: "Streptomycin is equally effective
and has less auditory toxicity than dihydrostreptomycin, and has
therefore essentially supplanted it for use in tuberculosis . . . If
vestibular toxicity to streptomycin occurs the dosage of the drug
can be reduced to 0.5 gms. per day without dropping the patient
from the protocol study; however, the substitution of dihydrostreptomycin would require him to be dropped. It is our belief
that this attitude is improper, especially in aged patients . . .
The vestibular toxicity of streptomycin can be extreme in patients who are in their 60's or 70's, and men in this age group
represent the single largest group at our hospital. The dizziness
and vertigo with movement develop promptly, and may be so
severe that the patient refuses to get out of bed. Hypostatic
pneumonia may develop. These older men frequently arise during the night to go to the bathroom; they often fall, in the dark,
as their vestibules no longer function properly . . . [The] auditory toxicity of dihydrostreptomycin, when it occurs, is tolerated

---

[1] A chemical derivative of streptomycin and at one time believed to be
less toxic, especially as regards vestibular disturbance.

by them much better than the early streptomycin toxicity. It is our practice to prefer dihydrostreptomycin, when use of either of these agents is necessary, in patients in this age group. They are watched carefully so that extremely prolonged therapy can be avoided if possible."

According to Etienne Bernard,[2] a tendency, prevailing in the 1950's, to use INH and PAS for the general treatment of tuberculosis and reserve streptomycin for special cases has recently given place to the use of all three drugs at the very beginning. After six weeks, when the antibiogram—the response of the particular culture of the organism to the various drugs—comes in, there should be a change in the treatment, using two of the major antibiotics to which the culture has remained sensitive. The age of the patient must, of course, be considered in setting up the particular regimen. Hinshaw and Garland (1963) also advised the use of the three drugs.

Bernard (1962) differentiates between regimens for the developed countries, on the one hand, and the developing, on the other. Among developed countries, he says, we may now expect the early eradication of tuberculosis in the United States, Canada, Holland, Denmark, whereas one may speak of control, but not of eradication, in France, England, and Sweden, though the latter two are beginning to approach the first group. In the developing countries of Africa, Asia, and South America, tuberculosis is still the most important infectious disease, and here other measures besides chemotherapy are required, notably BCG vaccination.

In developed countries, advances in therapy during the past few years have been so striking that tuberculosis has become a minor instead of a major disease. This is well illustrated in tables 5 and 6 and in figures 15-17. Before the advent of chemotherapy, the expectation of life for a person with newly diagnosed pulmonary tuberculosis averaged, in the absence of treatment (i.e., bed rest or collapse therapy), less than two years. At the present time, such persons can expect to live and, in most cases, to

[2]Personal communication.

achieve a normal life span. This changed situation has resulted in an increase in the proportion of the population who, at some time or other, have had active pulmonary lesions and have been restored to at least some degree of normal activity. The Army and Navy of the United States now return men to full duty after treatment for tuberculosis.

In summarizing the present status of the chemotherapy of tuberculosis, as well as hospital versus ambulatory treatment, Mc-Dermott (1960) stated that chemotherapy should be administered for eighteen to twenty-four months: "It is wise to continue drugs well past the time when most of the healing takes place . . . [The decision]whether or not to keep the patient in the hospital must be made primarily on the basis of the nature of the lesion and the conditions surrounding the patient. If possible, patients with highly infectious lesions, such as laryngeal disease, should be kept in the hospital, while patients with chronic cavitary lesions need not be kept in the hospital indefinitely unless the household associates would be at great risk." It is stated that the introduction of INH in the treatment of tuberculosis resulted in "divorcing the patient from the hospital" and in a lessening of surgery to remove tuberculous lesions. Although it is agreed that the most effective treatment for hospitalized patients is a combination of streptomycin with isoniazid, ambulatory patients using isoniazid alone or with PAS are said to be easier to keep under supervision and effective treatment. Some clinicians, however, are absolutely opposed to the treatment of active communicable tuberculosis with INH alone and on an ambulatory basis.

Augier (1961) summarized drug susceptibility studies of 464 patients with pulmonary tuberculosis admitted to the Bligny Sanatorium in 1959 and 1960 as follows: "Sixteen per cent of the patients who had not received antituberculosis treatment prior to admission showed primary resistance to 1 or 2 major antituberculosis drugs; 21 per cent of all the patients were admitted after receiving very irregular and inadequate chemotherapy and of those 80 per cent showed resistance to 1 or 2 major drugs. In 63 per cent of all patients having received adequate chemotherapy the results were similar during the first months of sanatorium stay, but after the fifth month results were excellent if streptomycin was given daily for the first 3 months; in these patients there were no cultures positive for tubercle bacilli at 7 to 8 months, whereas if streptomycin was given every other day 10 patients were still positive at 1 year, showing 100 per cent isoniazid resistance and 60 per cent streptomycin resistance. The necessity of triple therapy is stressed with isoniazid-PAS-streptomycin (the latter drug daily for at least 3 months)."

Sjöberg and Tivenius (1961) described 85 consecutive cases of patients with pulmonary tuberculosis who underwent thoracoplasty in a Swedish hospital during the years 1952-1954. All had triple drug therapy. "The over-all operative mortality rate during the period under review was 1.2 per cent. At the close of the observation period, 5 to 7 years after surgery, 82 of the 85 cases were considered arrested. Treatment failure was dependent partly on preoperative bacterial conditions and partly on the extent of the tuberculous process and cavity size." The conclusion was reached that "thoracoplasty still justifies its place among the surgical methods of treatment for pulmonary tuberculosis."

Crofton (1960) stated that if proper combinations of standard drugs were given, drug-resistant organisms would almost never emerge: "Sputum conversion to negative should always be achieved except in the very rare cases unfortunate enough to have been primarily infected with tubercle bacilli resistant to two or more of the standard drugs." When earlier, less effective drug combinations have been used, some patients may carry drug-resistant organisms. Such a patient "should be regarded as a challenge and meticulous resistance tests should be carried out on a number of cultures and no potentially useful drug should ever be risked until it is certain both that his organisms are sensitive to it and that it is given in combination with other drugs which have the best chance of being effective in preventing the growth of resistant mutants."

That antibiotics, combined with other chemotherapeutic agents, have the ability to control tuberculosis has been established in several countries. The appearance of fresh cases of tuberculosis in these countries has been gradually reduced, even while some patients have failed to respond to the new therapy because of inadequate early treatment or infection with resistant organisms. This was recognized by Florey (1961), who stated: "The sharp fall in the numbers of fresh cases of tuberculosis in England and Wales between 1953 and 1959 is another indication that tuberculosis is on the way to being controlled."

Resulting Social Problems

Unfortunately, the saving of many lives by chemotherapy has not always resulted in a change in the public attitude toward the disease and toward its victims. Tellesson (1960) stated that "A feeling of caution still prevails, and the individuals thus recovered may still suffer a social stigma, namely their non-acceptability in most phases of activity . . . A recent survey has shown that most citizens harbour the wildest misconceptions on this subject, despite the intensive educational efforts of the various authorities. The medical profession can help impress upon employer and employee (and the greatest resistance often comes from workmates unwilling to accept a known tuberculous colleague) that the man or woman who has had tuberculosis, who has been treated, and whose return to work is approved medically, can be a useful worker and constitutes no undue risk to those who work with him."

Another social and public health problem resulting from the saving of human lives from tuberculosis is the continuing hazard to the community from patients whose lesions have been brought under some degree of control, but whose sputum remains a source of infection for months and even years. These individuals usually harbor bacteria resistant to all major antimicrobial drugs, so there is an additional special reason for not wanting their infections to be spread. Unfortunately, the person who has a poor therapeutic result is often one who did not follow the advice of his physician at the beginning of treatment and is not likely to stay under any restrictions unless forced to do so. These people, many of whom are alcoholics, leave hospitals against the advice of their physicians, fail to attend clinics, and fail to take their antimicrobial drugs as prescribed. They thereby create a special problem, the seriousness of which has led to the establishment of detention hospitals where these patients can be kept until they are no longer infectious. In practice, the detention hospital is used only for care of the flagrant case constituting an immediate danger to other people. Nevertheless, the irresponsible person with chronic relapsing tuberculosis is a continuing double hazard because he has the potential of passing on infections resistant to the known effective drugs. Such persons are comparable to typhoid carriers and have been called "tuberculosis Marys," by superficial analogy with the case of "typhoid Mary," who was the source of so many typhoid epidemics. They are partly responsible for the fear of former tuberculosis victims, where there should be no fear.

Riley *et al*. (1962) have shown that some forms of tuberculosis, particularly of the larynx, may be even more contagious than measles. By channeling air from a specially equipped experimental tuberculosis ward in the Baltimore Veterans Administration hospital to an animal exposure chamber, aerial dissemination of pulmonary tuberculosis was measured. An average child with measles was found to produce about eighteen infectious units of the air-borne measles per hour, whereas the patient with tuberculosis laryngitis produced about sixty infectious units of air-borne tuberculosis in the same period of time. A small number of patients designated as "tuberculosis disseminators" were found to be responsible for the bulk of the animal infections. Treatment of the patients with antituberculous drugs dramatically reduced their infectiousness early in the course of treatment, even before any suppression of the numbers of tubercle bacteria in their sputum could be detected.

Thus the prevailing methods of treatment of patients suffering from tuberculosis present a variety of special problems, both clinical and sociological in nature. It is sufficient to mention hospital versus home care, bed rest versus ambulation; the proper use of therapeutic agents, especially for those who previously have not been treated; the duration of therapy; and the unsuccessful treatment of certain patients, resulting in cases of drug resistance. Thoracic surgery has been showing a steady decline, partly because of the reduction in the number of patients and partly because more effective treatment has been found to decrease the necessity of surgical intervention; clinicians have now far more confidence in chemotherapy to do the job alone.

In speaking of present-day problems in the eradication of tuberculosis, Soper (1962) pointed out that tuberculosis was at first a purely local problem. When eradication became possible, it took on national characteristics and had to be planned for on a national basis: "The decline of tuberculosis during the past decade and a half surpasses anything one could have imagined at the beginning of this period, and this with a minimal adaptation of preventive measures to the possibilities of the new era of chemotherapeutic cure and prevention of tuberculosis . . . There are already areas in the United States in which the first stage of eradication has been accomplished, viz., the interruption of transmission to the younger generation . . . The only serious difficulty of chemotherapy of tuberculosis is the necessity of continued medication over a long period of time . . . One cannot but ask what we are waiting for, what more must we have, what sign do we still require before assuming the responsibility for the eradication of tuberculosis? One does not have to see the end to start; but one must start if one is to see the end." Soper further called attention to the psychological significance of the transition from control, the objective of which is to reduce the incidence of tuberculosis to a low level and maintain it at that point, to eradication, which looks to complete elimination.

According to Canetti (1962b), "The possibility of eradicating tuberculosis in a country is essentially a function of its economic level. There is a fundamental difference in this respect between the countries called 'developed' from those called 'developing.'

. . . In the 'developing' countries, the practice of chemotherapy meets enormous difficulties. The principal ones are the cost of drugs usually given with isoniazid and the tendency of patients to stop treatment too soon. The finding of chemotherapeutic regimes which are really applicable in the conditions of 'developing' countries, and methods of assuring regular treatment and effective self-medication are two problems of fundamental importance. These are not yet resolved."

MacGregor (1962) reminds us that there are other important aspects of the control and eventual eradication of tuberculosis: "Because treatment is now so highly effective I regard tuberculosis case-finding as the key to its elimination. The search for new patients must be pursued with the utmost vigour and must employ progressively more refined techniques based on a more intimate knowledge of the epidemiological pattern of the disease . . . We have come to the stage when every death should be investigated and the source of every new case identified wherever possible."

It may be added that Weber *et al.* (1963) in a long-term study of 137 patients, aged fifty years and older, found "no evidence to suggest that there is any inherent difference in the response of elderly subjects to chemotherapy of tuberculous disease compared with that found in younger age groups . . ."

The problem of tuberculosis is both global in scope and still the problem of every general practitioner of medicine. In fact, a transition is now taking place from the care of tuberculosis patients in the specialized clinic and sanatorium to care in the office of the private physician. This shift is more evident in Europe than in the United States.

In France (Bernard, 1962) mortality from tuberculosis was receding at a rate of 3 to 5 per cent annually over the fifteen-year-period preceding 1952, but in that year, the year isoniazid was introduced, the reduction was 27.8 per cent. Tuberculous meningitis, a disease heretofore nearly always fatal, yielded to streptomycin with a survival rate of 30 to 45 per cent. The introduction of INH pushed the recovery rate to 90 per cent. But it was the combination of these drugs with the addition of PAS that produced the best results and prevented the emergence of resistant organisms.

In a recent analysis of new active cases of tuberculosis in large cities as compared with the country as a whole, the U.S. Public Health Service reported (Tuberculosis Chart Series, 1961) that cities of 500,000 population and above showed a case rate of 64.0 per 100,000; cities of 100,000 to 500,000 showed a case rate of 37.7; and all other areas showed a case rate of only 25.1. The case rate for the United States as a whole was 32.5 per 100,000 (table 11). The rates of new cases of tuberculosis in various countries are given in table 12.

TABLE 11.
New Active Cases of Tuberculosis in the United States, 1959

Age, years	Number	Per cent of total	Rate[a]
Under 25	9,603	16.7	12.3
25-44	18,364	31.9	39.1
45-64	19,063	33.1	53.5
65 and over	10,505	18.3	64.7
All ages	57,535	100.0	32.5

[a]Per 100,000.
Source of Data: *Tuberculosis Chart Series* (1961c).

TABLE 12.
New Cases of Tuberculosis Registered in Selected Countries, 1959-1961

Country	Case rate per 100,000 population		
	1959	1960	1961
United States	32.5	30.8	29.3
Puerto Rico	106.2	90.4	82.6
Canada	36.9	35.5	32.7
Peru	388.9	376.7	365.7
Denmark	26.1	24.7	...
Finland	157.3	171.0	182.2
Netherlands	59.2	50.9	46.0
Czechoslovakia	176.0	132.7	103.7
Japan	537.7	524.2	445.9

Source of Data: Lowell (1962).

Using as a base the new active tuberculosis case rates for the years 1952-1959, the Public Health Service, through its committee on Goals and Standards in Tuberculosis Control, thought that it should be possible to achieve a new active case rate of not more than 10 per 100,000 population by 1970. (See figs. 18-20).

The tuberculosis situation in Europe is far from encouraging, however, and is "disappointing and depressing . . . serious and even dangerous," according to an editorial in the *Bulletin* of the International Union against Tuberculosis (1962): ". . . . only small population groups in Europe are covered by a tuberculosis programme in which our existing knowledge and possibilities are rationally applied. Even if the epidemiological situation today is quite different from what it was in the early part of this century, the organization of the tuberculosis programme seems, in many countries, to have not much changed during the last decades. [There are] far too few young doctors . . . going into the specialty of tuberculosis and you cannot blame them, when they are told that tuberculosis is no longer a problem . . . When the politicians and the administrators responsible for allocating the money for health programmes are also repeatedly told that tuberculosis is no longer a public health problem, they naturally tend to cut down the allocation for the tuberculosis programme and give the money to cancer and heart programmes . . . "

Although it is agreed that antituberculous drugs when used properly are nearly 100 per cent effective, "Much harm has been done and is still being done by publishing mortality figures for tuberculosis and giving the impression that the decline in tuberculosis mortality can be taken as an expression of the decline in the tuberculosis problem," said the *Bulletin,* which went on to urge that "we should stop publishing the figures for tuberculosis mortality and giving the public to understand that by a reduction in these figures, we indicate that the tuberculosis problem is almost under control"

Lowell (1962) also takes a realistic view. In the 1961 report on tuberculosis in New York City he states that "all indices reflecting the prevailing trend of tuberculosis in New York City point to measurable gains and improvements made during 1961. However, these accomplishments should not obscure the fact that each year thousands of people continue to be newly infected with tuberculosis and fall prey to this debilitating disease." The conclusion was reached that "Tuberculosis remains New York City's most challenging and costly single public health problem."

No attempt can be made here to recommend various regimens for the treatment of the different manifestations of tuberculosis. Therapy has to depend on the nature and state of the disease, the age of the patient and his previous treatment, the economic and family status of the patient, the economy of the country, the facilities available for treatment, the nature and sensitivity of the organism responsible for the particular case, and numerous other factors. These are considerations to be weighed by the physician when he decides on a regimen, whether it will be triple- or double-drug or single-drug therapy, what drugs will be used and in what dosage. The introduction of additional antituberculous drugs, subsequent to streptomycin in 1944, PAS in 1947, and isoniazid in 1952, increased the physician's armamentarium. Although these drugs are useful in special circumstances, as is surgery, the most important and effective drugs continue to be streptomycin, isoniazid, and PAS.

BCG Vaccinations and Chemoprophylaxis

The advisability of using BCG depends on the particular need for it in certain individuals and in certain population groups. It should by no means be considered as a substitute for other control measures, but rather used in addition to these, when justified. As far as the United States is concerned, BCG is not needed as a general measure and is even counterindicated by the fact that it interferes with the use of the tuberculin test as an epidemiologic and diagnostic tool (Long et al., 1957).

The importance of BCG vaccine in controlling tuberculosis

among the Indians in the United States, including Alaska, was evaluated by Aronson and Aronson (1952). Among a total of 1,551 vaccinated persons, over a period of years, twelve died from tuberculosis, giving a rate of 0.56 per thousand per year of observation. Among the 1,457 non-vaccinated persons who served as controls, sixty-five died of tuberculosis, at a rate of 3.32 per thousand per year.

Hobby (1962b) makes the additional point that BCG vaccination never succeeds in completely preventing the development of tuberculosis, the degree of immunity produced by BCG being at best no greater than that produced by natural infection which results in a positive reaction to tuberculin. "The duration of immunity produced by BCG vaccination is uncertain, and there are many strains of BCG which vary widely in their immunizing qualities . . . The tuberculin test is such a useful means of detecting infection with virulent tubercle bacilli that one cannot afford to employ, on a mass basis, any substance which interferes with the use of this diagnostic tool in countries where the prevalence of tuberculous disease is low and the need for accurate diagnosis is great."

The above does not hold, however, for many large areas of the world, where tuberculosis is still a huge problem. Here, BCG vaccination has achieved tremendous results, including the practical suppression of tuberculous meningitis, where this complication is exceedingly rare.

Chemoprophylaxis made its appearance in 1958. It was recommended too that the practice be begun in childhood and adolescence and that each day, for a period of about eight months, INH be taken alone or in combination with PAS. In most cases, the treatment was limited to children under four years of age, adolescents at the puberty stage, young people living in a contaminated environment, and those that gave a strongly positive tuberculin test. It was also often used as a prophylaxis against relapse in certain cases. For further reference see Zorini (1962).

Canetti (1962b) has advised: "The desire to see tuberculosis disappear as rapidly as possible should not lead to the application of excessive measures, measures that are disproportionate to the danger which the disease represents at present in a particular country. This remark concerns in particular certain measures of active prophylaxis (chemoprophylaxis) which are likely to be applied on a large scale. The number of theoretically justified prophylactic measures of all kinds is continually increasing in medicine; some of these are applied effectively. However, the sum of all the measures of this kind, the 'prophylactic stress,' as it were, under which a healthy population can be kept, is limited."

THE FINAL CHAPTER

In considering the status of tuberculosis, Long wrote in 1953: "Altogether, it is a period of intensive treatment. Sanatoria designated as such, have virtually disappeared. Institutions for the treatment of tuberculosis are now designated and rightly so as 'tuberculosis hospitals.' A few weeks ago I saw the change pithily epitomized in a sentence or two by a keen lay student, Thomas Mann, who was looking back at *The Magic Mountain* thirty years after he created it. He wrote: 'Such institutions as the Berghof were a typical pre-war phenomenon. . . . *The Magic Mountain* became the swan song of that form of existence. Perhaps it is a general rule that epics descriptive of some particular phase of life tend to appear as it nears its end. The treatment of tuberculosis has entered on a different phase today; and most of the Swiss sanatoria have become sports hotels.' "

Speaking in 1954 before the Veterans Administration–Army-Navy conference on the chemotherapy of tuberculosis, J. L. Wilson suggested that: "In the world one hundred years hence, tuberculosis will probably be as much a rarity as leprosy is today. Medical students and physicians taking postgraduate courses will learn the textbook picture of tuberculosis and be told that they must always be on the watch for cases arising in endemic foci in certain backward parts of the country or flown in from remote places of the shrunken globe."

Lowell (1958) in a discussion of a "search for tuberculosis in a transition period," suggested *ad astra per aspera* as a motto to guide the antituberculosis movement toward the final eradication of the disease—"a goal, that some of us may see attained, when the White Plague will be more or less of a medical curiosity rather than the relative scourge it is today"—and as a warning lest the "routine everyday tasks in controlling tuberculosis may dull our sense of achievement and vision so that there can be a lessening of the enthusiasm that is so sorely needed in fighting a disease of chronic character." Lowell added: "We must retain a sense of dedication to the cause which is as urgent today as it was in the time of Trudeau. The ultimate objective should be constantly in the mind's eye in the midst of our daily work."

Five years later, Drolet and Lowell (1962) spoke of *the last stage,* citing data as of 1959 which showed that the general tuberculosis death rate for all ages was 5.5 per 100,000 in Canada, 6.5 in the United States,[3] 8.4 in England and Wales, and 23.4 in France. Since then, the mortality rates have dropped even further in these and other countries.

Yeager (1962) stated in testimony before United States Senate Health Appropriations Committee: "We, ourselves, must take the blame if the eradication of tuberculosis is not achieved within the forseeable future in the United States." An anonymous reviewer, in a World Health Organization report (1962), has come to the conclusion that research on tuberculosis has progressed to such an extent that "It may soon be feasible to place in the hands of the non-literate villager more real power to affect the outcome of a tuberculous infection than could have been exerted by the most highly trained physician of little more than a decade ago." The degree of control of the disease in technically advanced countries is such that "complete eradication appears a distinct possibility." Emphasis should be put upon improving the diagnostic tools: "Tuberculosis differs essentially from such great killing diseases of today as cancer and atherosclerosis in that there

[3]This rate dropped in 1962 to 4.9.

are no mysteries about its causation, diagnosis and treatment
. . . The use of the new drugs in different environments; the
habits and traditions affecting the personal responsibility of patients for continuing treatment; a better understanding of the
epidemiology and immunological aspects of the disease: these are
the principal factors involved in the large-scale application of
modern tuberculosis chemotherapy."

In addressing the Sixteenth International Conference on Tuberculosis, held in Canada in 1961, Canetti (1962*a*) as summarized
by the chairman stated: "We are not concerned with *eradication* in the absolutely literal sense of the word because this
is something that many of us believe to be biologically impossible. What we are talking about, however, is tuberculosis
'eradication' in the sense of reducing the problem to the point
where the disease is a scientific curiosity. This *is* a biologic
possibility. Moreover, it is an attainable, practical goal because
it has already happened in fairly large geographic areas that
are well populated [and therefore] it is permissible for us
to think in terms of tuberculosis 'eradication' for the economically
developed areas as very much the same sort of problems are
involved in all of these areas. It is likewise appropriate for us to
think in terms of some sort of tuberculosis *control* for the economically underdeveloped areas. Moreover, it is appropriate for
us to consider both of these things at the same time (tuberculosis
eradication and tuberculosis control) because many of the steps
involved with today's technology are very much the same in the
two widely different types of situation."

In speaking of the present status (1962) of tuberculosis control
and eradication, Brightman and Hilleboe said: "The white
plague, tuberculosis—is retreating in America. The decade ahead
of us, the sixties, will be decisive. We are determined . . . that
the retreat of the tubercle bacillus shall inflict as few casualties
upon our human resources as possible. [Today] tuberculosis
workers have found it increasingly more difficult to find the persons who have active tuberculosis [and] control work requires
case-finding methods [that] have the accuracy of high-powered

rifles. . . . [If] we work hard in the decade ahead, tuberculosis is one disease that we can relegate to a position of minor importance in public health . . .

"The developments in chemotherapy and thoracic surgery have altered the management of tuberculosis from a leisurely pace based on good supportive care and watchful waiting to a dynamic approach with specific therapies capable of producing early control of infectiousness and an arrest of the disease, usually within a predictable and comparatively short period of time. The 1957-1959 average rate for new active cases of tuberculosis was 36.9 per 100,000 for the country as a whole, compared with 76.4 in 1940. The mortality rates have shown the same trend with tuberculosis dropping from its position as the leading cause of death in 1900 to eighth place in 1940 and eleventh place in 1960. The proportion of tuberculin reactors is steadily falling at all age levels, particularly in children and young adults. The special tuberculosis hospitals, both public and private, have been closing at a rather rapid rate and the operating institutions are showing lower and lower occupancy rates.

"The changing epidemiological pattern of tuberculosis calls for a reevaluation of control measures and a reorientation of public health measures if eventual eradication or at least reduction to the status of a rare disease is to be achieved. The public health agency's responsibility in tuberculosis may be expected to increase rather than decrease as periods of hospitalization for active tuberculosis shorten and an increasing proportion of cases are not hospitalized at all. [Thus] the future may well call for the establishment of tuberculosis wards as parts of general hospitals for the care of patients during the early weeks or months of intensive chemotherapy, with convalescent homes or organized home care programs providing care thereafter."

Outbreaks, especially those of an epidemiological nature, may still occur. This is true where individuals are placed in closed compartments, as on ships, for any long period of time (Ochs, 1962). But measures are available for preventing such outbreaks, for diagnosing them when they occur, and finally for curing the afflicted individuals.

Finally, the present status of tuberculosis control can be glimpsed from the *Transactions of the 20th Research Conference in Pulmonary Diseases* (1961*b*). This publication, assuming a third title in its fifteenth year, has thus reflected the rapid changes that have taken place since the first of these conferences was held in December, 1946. It was first designated as the *Minutes of the Streptomycin Conferences.* After nine such conferences were held semiannually, the title was changed in 1950 to *Transactions on the Chemotherapy of Tuberculosis,* the word "streptomycin" being dropped from the title in view of the advent of additional chemotherapy agents. These conferences were then held annually. After another ten years of progress, the name of the conference was changed again, to *Pulmonary Disease Conferences,* the words "Chemotherapy of Tuberculosis" being dropped.

The first conference, held in 1946, had 56 registered participants, and the reports ran to 82 typed pages. The growth of these conferences reached a peak in 1957, with 447 participants and a volume of *Transactions* running to 616 pages. Subsequently the registrations at the conferences began to decline; this reflected a decrease in tuberculosis patients and in the number of physicians concerned with tuberculosis, as well as the shift of emphasis from tuberculosis to nontuberculous pulmonary diseases.

THE REMAINING PROBLEMS

Tuberculosis is becoming more and more a disease of the older ages. It has been reported (*Tuberculosis Chart Series,* 1958) that the median age for newly reported active cases for the United States is approximately forty-four years; that is, half of the newly reported cases are under 44 and those of the other half are over 44 years. When examined by race and sex, the new active case rates show about the same distribution as the corresponding death rates. Nonwhite males have the highest case and death rates of all race and sex groups—about three times that of white males. The high ratio of cases per death is largely owing to the decline of tuberculosis deaths during recent years, accompanied by only a relatively small decline in the number of tuberculosis cases reported.

Although we can speak optimistically of the long-range prospects for the world-wide conquest of tuberculosis, there are wide disparities in the remaining problem geographically. Those most advanced in their tuberculosis control programs are the countries in northern Europe, North America, and the down-under countries of Australia and New Zealand. The disease is still a major cause of illness and death in most of Asia, Africa, and South America. The International Union against Tuberculosis recently reported (1962) an estimate of 30 cases of active tuberculosis per 1,000 in these countries. In northern Europe the figure would be nearer 1 per 1,000.

In the more advanced countries children now show very low rates of sensitivity to the tuberculin test. In the United States only about 5 per cent of the recruits going into the Navy have positive tuberculin tests. At this age, about nineteen years, almost all young men in the developing countries have positive tests, indicating a continuing serious spread of the infection.

In India, according to Prasad (1961), "Since the introduction of a malaria eradication program, pulmonary tuberculosis has become the chief cause of death . . ." A sample survey undertaken between 1955 and 1958 was the first major attempt at a systematic assessment. "The morbidity rate revealed by this survey was 13 to 25 per 1,000 of the population. The survey also revealed that there is not much difference in the prevalence rates in urban and rural areas. In industrial cities about 75 per cent of the population become tuberculin positive by the age of 15. It is estimated that in India there are about 2.5 million open cases of tuberculosis and approximately a half million people die annually from this disease. The mortality rate from tuberculosis is about 150 per 100,000 of the population . . . It has been found that treatment of patients in their homes with antibacterial drugs is as effective as treatment in hospitals and sanatoriums."

The present situation in the United States has been discussed by Yeager (1962), who called attention to the fact that "tuberculosis is still largely a disease of the lower economic groups and particularly of older people in these groups." He adds: "These individuals have multiple illnesses, not simply tuberculosis. Facilities must, therefore, be available where adequate consultation can be obtained and adequate studies made of these coexisting illnesses . . . Hospitalization is still both advisable and necessary in the treatment of many cases with active tuberculosis. The length of time necessary for a patient to remain hospitalized is being rapidly reduced [and] when adequate antimicrobial therapy is used it is not necessary to use complete bed rest in the majority of the cases. The average length of stay is shortened so that many conservative institutions are discharging minimal cases at an average of 3 or 4 months, and moderately advanced, or early far advanced cases only a few months later."

At the fifteenth conference of the International Union against Tuberculosis, held in Istanbul in 1959, the National Tuberculosis Association of the United States proposed a resolution dealing with the "eradication of tuberculosis throughout the world as a public health problem." It was stipulated that the word "eradication" should mean "the ultimate goal of having no human being on this earth react significantly to a skin test with a proper dosage of a standardized tuberculin; in other words, total bacteriological eradication so far as tuberculosis is concerned, although there may well be established intermediate goals to strive for, step by step, before arriving at this long range objective." There are, the Association stated further, "many small areas today where tuberculosis has been eradicated," and there is "no reason to assume it is inevitable that tuberculosis will appear again in those areas, or if it does, that it cannot be contained and eliminated again and rather rapidly. . . . Effective tools for detection, treatment and prevention of tuberculosis are now in existence which never were available in centuries past to attack this ancient scourge of mankind, which today is probably the most important communicable disease in the world as a whole."

On the other hand, an expert commmmittee of the World Health Organization (1960) has held that tuberculosis should be regarded as a public health problem "until the prevalence of natural (and specific) reactors to tuberculin among children in the 14 year age group has become less than 1 per cent." Commenting on this, a correspondent of *Lancet* (May 26, 1962) remarked that "on this basis, even in affluent and advanced countries, talk of eradication of the disease sounds premature, while in the newly independent African countries more than a generation must pass before the infection becomes anything like so rare."

The optimistic view, on the part of the general public and even of some physicians, that tuberculosis has already been controlled, both as a disease and as a social danger, has not been generally accepted. Although it is generally agreed that therapeutic advances have been remarkable during the last fifteen years and that mortality from tuberculosis has been considerably reduced, the fact that the curves for morbidity do not follow in a parallel way makes some people pause and question the optimism which leads some to speak of the forthcoming complete eradication of tuberculosis. It is, of course, necessary to keep in mind the fact that the insufficient reduction in morbidity may be more apparent than real, since, with the advances in methods for detecting infectivity, many more cases are now detected than was possible previously (Puig, 1963). The conclusion may be reached that none of the classical means of treatment should be neglected, especially the use of dispensaries, hospitalization, and general hygienic measures. Prophylactic surveys and vaccination should also be emphasized, and the public is to be advised that tuberculosis is a disease which has not yet ceased to be a serious social and economical medical problem (Guasch, 1963).

Holm (1962) believes that tuberculosis eradication is no longer a theoretical problem and has become a realistic possibility. He calls particular attention to the distinction in concept between *control* and *eradication* of this disease: "The elimination of tuberculosis as a major public health problem in a population has been achieved when less than 1 per cent of the children of fourteen years of age are specific reactors to the WHO standard tuberculin test." The Netherlands, Denmark, several states in the U.S.A. and provinces in Canada, and certain other countries are now making definite plans for the eradication of tuberculosis. "At the present time," Holm concludes, "no country has achieved complete control of tuberculosis, but this stage will be reached shortly. It is therefore high time to plan an eradication programme."

EPILOGUE

As I look back over the past twenty years, the period covering the isolation of streptomycin and other antibiotics, the life-saving drugs produced by microbes, and, further, as I look back over the more than four decades I have devoted to the study of microbes in the soil and in the sea, I can see a marvelous tapestry unfolding. In this tapestry are woven thousands of strands of gold and silver discovered by the microbiologist and the chemist, the physiologist and the pharmacologist, and finally the clinician —all searchers of the unknown, who have combined their efforts to produce this tapestry. The saving in human lives has been untold; the life span of the average human being has been increased by two decades; numerous diseases have been brought under complete control. But most important, the ancient foe of man, known as consumption, the great white plague, tuberculosis, or by whatever other name, is on the way to being reduced to a minor ailment of man. The future appears bright indeed, and the complete eradication of this disease is in sight.

One may well be permitted to agree with the editorial writer of the *Chest and Heart Bulletin* (1962a) who attempted to summarize the factors responsible for the present control of tuberculosis in Western Europe and North America:

"Some will give all credit to our three modern drugs, now used with increasing confidence. They will claim that through this means alone, tuberculosis could be wiped out from Asia and Africa. Others argue that hygiene, in the broader sense of good nutrition, housing and freedom from strain, have been more important, and that drugs represent merely one factor in a campaign of social progress. In both arguments there is truth. Neither method can be relaxed in the hope that its counterpart would do all the work. We need more of the drugs—well used. And Asia and Africa need food and houses and less toil. In these eighty years, a great saga of medicine has been sung, full of despair and hope. Thousands of tuberculosis invalids have sought pine forests among the hill tops, and have wrestled with the peculiar depression which tuberculosis brings. Often the doctors who looked after them were tuberculous themselves. In no disease have so many remedies—medical, surgical and social—been tried —mostly with limited success—until the new drugs arrived."

BIBLIOGRAPHY

Anonymous

1958. *Tuberculosis Chart Series.* Washington, D.C.: U.S. Department of Health, Education and Welfare, Public Health Service.

1960. *World Health Organization Expert Committee on Tuberculosis: 7th Report.* Tech. Rept. No. 195.

1961a. *The Glaxo Volume,* No. 22, 21-28. Middlesex, England: Glaxo Labs., Ltd.

1961b. *Transactions of the 20th Research Conference in Pulmonary Diseases.* Washington, D.C.: Veterans Administration, Dept. of Medicine and Surgery.

1961c. *Tuberculosis Chart Series.* Washington, D.C.: U.S. Department of Health, Education and Welfare, Public Health Service.

1962a. *Chest and Health Bulletin* (editorial), Dec., 1962, pp. 176-177. London: The Chest and Heart Association.

1962b. "Tuberculosis Research" (editorial), *World Health Organization Chronicle,* 16:42-45.

1962c. *World Health Organization Epidemiological and Vital Statistics Report* (Geneva), 15:327-388.

1963. *Quarterly Progress Report of the Veterans Administra-*
        tion—Armed Forces Study on Chemotherapy of Tuber-
        culosis, April. Washington, D.C.: Veterans Administra-
        tion, Dept. of Medicine and Surgery.
Amberson, J. B., Jr., McMahon, B. T., and Pinner, M.
    1931. "A Clinical Trial of Sanocrysin in Pulmonary Tuber-
        culosis," *Amer. Rev. Tuberc.,* 24:401-435.
Aronson, J. D., and Aronson, C. F.
    1952. "Appraisal of Protective Value of BCG Vaccine," *Jour.*
        Amer. Med. Assoc., 149:334-343.
Ashburn, P. M.
    1947. *The Ranks of Death.* New York: Coward-McCann.
Augier, J.
    1961. "Results and Interpretations of Cultures and Drug Sus-
        ceptibility Depending on Treatment Before Admission to
        the Santorium (A Statistical Study of 464 Cases of Pul-
        monary Tuberculosis)," *Rev. Tuberc.* (Paris), 25:590-617.
Babcock, R. H.
    1907. *Diseases of the Lungs.* New York: D. Appleton.
Babès, V.
    1889. "Sur les associations bactériennes de la tuberculose," in
        Congrès pour l'Etude de la Tuberculose, 1888, (Paris),
        pp. 542-56.
Baldwin, E. R.
    1913. "Tuberculosis: History and Etiology," in W. Osler and
        T. McCrae, *Modern Medicine,* 2d. ed., I, 287-338. New
        York: Lea & Febriger.
Barber, M.
    1951. "Medical Services in the Soviet Union," *Lancet,* 261:775-
        778.
Barnett, T. B., *et al.*
    1960. "The Chemotherapeutic Management of Treatment Fail-
        ures and Relapses in Pulmonary Tuberculosis," *Amer.*
        Rev. Resp. Dis., 82:751-752.
Barnwell, J.
    1952. "Presentation of Trudeau Medal to Dr. Amberson," in
        National Tuberculosis Association, *Transactions, 48th*
        Annual Meeting, pp. 1-6.
Bentley, P.
    1948. *The Brontës.* Denver: Alan Swallow.

Bernard, E.
    1962. "Évolution de la chimiothérapie antibacillaire au cours
        des dix dernières années," *Acta Phtisiol.* (Paris), pp. 3-11.
Biggs, H. M.
    1888-1889. "The Principles of Treatment in Pulmonary Tuber-
        culosis, with Some Observations of Its Etiology," *Buf-
        falo Med. & Surg. Jour.*, 28:633.
Brightman, I. J., and Hilleboe, H. E.
    1962. "The Present Status of Tuberculosis Control," *Amer.
        Jour. Pub. Health*, 52:749-758.
Brown, C. A.
    1937. *Life of John Keats.* London: Oxford Univ. Press.
Brown, L.
    1941. *The Story of Clinical Pulmonary Tuberculosis.* Balti-
        more: Williams & Wilkins Co.
Brownlee, J.
    1918. *An Investigation into the Epidemiology of Phthisis in
        Great Britain and Ireland.* London: Medical Research
        Council.
Bunn, P. A.
    1947. *Streptomycin in the Treatment of Tuberculosis.* Wash-
        ington, D.C.: Conference on Antibiotic Research.
    1949. "Tuberculous Meningitis," *Natl. Tuberc. Assoc. Bull.*,
        October, p. 135.
Canby, H. S.
    1939. *Thoreau.* Boston: Houghton Mifflin Co.
Canetti, G.
    1962a. "Summary by the Chairman," *Bull. Intern. Union
        against Tuberc.*, 32:698-699.
    1962b. "The Eradication of Tuberculosis: Theoretical Problems
        and Practical Solutions," *Tubercle*, 43(3): 301-321.
Cantani, A.
    1885. "Tentativi di bacterioterapia," *Riforma Medica Napoli*,
        p. 147.
Castiglioni, A.
    1933. *History of Tuberculosis.* New York: Medical Life Press.
Chaves, A. D., *et al.*
    1961. "The Prevalence of Drug Resistance among Strains of
        M. tuberculosis Isolated from Ambulatory Patients in
        New York City," *Amer. Rev. Resp. Dis.*, 84:744.

Clark, P. F.
   1961. *Pioneer Microbiologists of America.* Madison: Univ. of Wisconsin Press.
Cocchi, C.
   1948. "Terapia della tuberculosi con streptomicina solfone e vitamina A (Tuberculosi miliare e meningite tubercolare)," *Riv. Clin. Ped.,* 46:1-67.
Cooke, R. E., Dunphy, D. L., and Blake, F. G.
   1946. "Streptomycin in Tuberculous Meningitis: Report of Its Use in One-Year-Old Infant," *Yale Jour. Biol. Med.,* 18: 221-226.
Cooper, D. A.
   1953. "Presidential Remarks," in National Tuberculosis Association, *Transactions, 49th Annual Meeting,* pp. 18-21.
Crofton, J.
   1960. "The Chemotherapy of Tuberculosis, with Special Reference to Patients Whose Bacilli Are Resistant to the Standard Drugs," *Brit. Med. Bull.,* 16:55-60.
Cummins, S. L.
   1949. *Tuberculosis in History.* London: Bailliere, Tindall and Cox.
Debré, R., and Brissaud, H. E.
   1953. *Meningite tuberculeuse et tuberculose miliaire de l'enfant.* Paris: Masson & Cie.
Drabkin, I. E.
   1950. *Caelius Aurelianus. De morbis chronicis. On Acute Diseases and on Chronic Diseases.* Chicago: Univ. of Chicago Press.
Drolet, G. J.
   1923. "100,000 Victims of Tuberculosis," *Bull. N.Y. Tuberc. Assoc.,* 4(3):17-21.
   1946. "Epidemiology of Tuberculosis," in B. Goldberg, ed., *Clinical Tuberculosis,* pp. A3-A70d. Philadelphia: F. A. Davis Co.
Drolet, G. J., and Lowell, A. M.
   1955. "Whereto Tuberculosis? The First Seven Years of the Antimicrobial Era, 1947-1953," *Amer. Rev. Tuberc.,* 72: 419-452.
   1962. "Tuberculosis Mortality among Children: The Last Stage," *Dis. of the Chest,* 42:364-371.

Dubos, R., and Dubos, J.
1952. *The White Plague: Tuberculosis, Man and Society*, Boston: Little, Brown & Co.
Ebstein, E.
1932. *Tuberkulose als Schicksal*. Stuttgart: Ferdinand Enke Verlag.
Ellis, A. E.
1959. *The Rack*. Boston: Little, Brown & Co.
Fanconi, G., and Löffler, W.
1948. *Streptomycin und Tuberkulose*. Basel: Benno Schwabe Verlag.
Feldman, W. H.
1954. "Streptomycin: Some Historical Aspects of Its Development as a Chemotherapeutic Agent in Tuberculosis," *Amer. Rev. Tuberc.*, 69:859-868.
Feldman, W. H., Hinshaw, H. C., and Mann, F. C.
1945. "Streptomycin in Experimental Tuberculosis," *Amer. Rev. Tuberc.*, 52:269-298.
Fishberg, M.
1932. *Pulmonary Tuberculosis*. 3d ed. New York: Lea & Febiger.
Flick, L. F.
1925. *Development of Our Knowledge of Tuberculosis*. Philadelphia: Wickersham Printing Co.
Florey, M. E.
1961. *The Clinical Application of Antibiotics*. Vol. II. *Streptomycin and Other Antibiotics Active against Tuberculosis*. London: Oxford Univ. Press.
Forlanini, C.
1894. "A contribuzione della terapia chirurgica della tisi ablazione del polmone? Pneumotorace artificiale?" *Gazz. Med. di Torino*, 45:381-384, 401-403.
Fouquet, J., *et al.*
1949. "Résultats de deux ans de traitement de la méningite tuberculeuse de l'enfant par la streptomycine au Centre de la Salpêtrière," *Bull. Mém Soc. Méd. Hôpitaux Paris*, 65:13-14, 553-557.
France, Anatole
1924. *The Latin Genius*. Trans. Wilfrid S. Jackson. New York: Dodd, Mead & Co.

Gratia, A.
  1948. "Lecons à tirer de la conférence interuniversitaire sur le traitement par la streptomycine des méningites tuberculeuses et de la granulie," *Rev. Méd. Liége,* 4:48-51.
Guasch, E. A.
  1963. "Evolución de la morbilidad tuberculosa, durante veinte años, vista desde un dispensario," *Inst. Antituberculoso "Francisco Moragas,"* 15:29-40.
Hale-White, W.
  1938. *Keats as Doctor and Patient.* New York and London: Oxford Univ. Press.
Hart, P. D.
  1946. "Chemotherapy of Tuberculosis Research during the Past 100 Years," *Brit. Med. Jour.,* 2:805-810.
Hill, L.
  1960. "The Conquest of Tuberculosis," *Amer. Rev. Resp. Dis.,* 82:291-294.
Hilleboe, H. E., and Brightman, I. J.
  1961. Editorial, "The Big Push Ahead in Tuberculosis Control," *Amer. Rev. Resp. Dis.,* 84:893-898.
Hinshaw, H. C.
  1954. "Historical Notes on Earliest Use of Streptomycin in Clinical Tuberculosis," *Amer. Rev. Tuberc.,* 70:9-14.
  1955. "Progress in Tuberculosis Therapy," *Jour. Amer. Med. Assoc.,* 158:832.
Hinshaw, H. C., Feldman, W. H., and Pfuetze, K. H.
  1946. "Treatment of Tuberculosis with Streptomycin," *Jour. Amer. Med. Assoc.,* 132:778-782.
Hinshaw, H. C., and Garland, L. H.
  1963. *Diseases of the Chest.* 2d ed. Philadelphia: W. B. Saunders Co.
Hirsch, A.
  1883. *Handbook of Geographical and Historical Pathology.* Trans. Charles Creighton. London: New Sydenham Society. Vol. III.
Hobby, G. L.
  1962a. "Prospects for a Vaccine against Tuberculosis," *NTA Bull.,* May, pp. 12-13.
  1962b. "Primary Drug Resistance in Tuberculosis," *Amer. Rev. Resp. Dis.,* 86:839-846.

Holm, J.
1962. "The Eradication of Tuberculosis," *Dis. of the Chest,*
February, pp. 18-21.
Hornung, S.
1962. "L'état actuel de l'endémie tuberculeuse et la question de
son éradication en Pologne: Le rôle de la législation anti-
tuberculeuse," *Bull. Intern. Union against Tuberc.,* 32:
683-688.
Huber, J. B.
1906. *Consumption: Its Relation to Man and His Civilization.*
Philadelphia: J. B. Lippincott Co.
Jacobson, A. C.
1926. *Genius: Some Revaluations.* New York: Adelphi Publications.
Keefer, C. S., *et al.*
1946. "Streptomycin in the Treatment of Infections: A Report
of One Thousand Cases," *Jour. Amer. Med. Assoc.,* 132:
4-11, 70-77.
Kervran, R.
1960. *Laënnec: His Life and Times.* Trans. D. C. Abrahams-
Curiel. New York and London: Pergamon Press.
Klebs, T. A. E.
1873. "Die künstliche Erzeugung der Tuberkulose," *Arch.
Exper. Pathol. Pharmakol.,* 1:163-180.
Koch, R.
1932. *The Aetiology of Tuberculosis.* New York: National
Tuberculosis Assoc.
Krafchik, L. L.
1946. "Tuberculous Meningitis Treated with Streptomycin,"
Jour. Amer. Med. Assoc., 132:375-376.
1948. Staff proceedings, St. Peter's General Hospital, New
Brunswick, N.J., July.
Krause, A. K.
1928. "Tuberculosis and Public Health," *Amer. Rev. Tuberc.,*
18:271-322.
Krause, A. K.
1932. "Introduction to the Aetiology of Tuberculosis," *Amer.
Rev. Tuberc.* 25:285-298.
Laënnec, R. T. H.
1819. *De l'auscultation médiate.* Tran. by H. Théophile. 2

vols. Paris: J. A. Brosson. London: J. Bale Sons & Daniel-
son, Ltd., 1923.

Lehmann, J.
1946. "Para-aminosalicylic Acid in the Treatment of Tuberculosis," *Lancet*, 251:15-16.
1949. "The Treatment of Tuberculosis in Sweden with Para-
aminosalicylic Acid (PAS): A Review," *Dis. of the
Chest.*, 16:684-703.

Lincoln, E. M.
1947. "Tuberculous Meningitis in Children," *Amer. Rev.
Tuberc.*, 56:75.
1961. "Eradication of Tuberculosis in Children," *Arch. En-
vironm. Health*, 3:444-455.

Lincoln, E. M., Sordillo, S. V. R., and Davies, P. A.
1960. "Tuberculous Meningitis in Children: A Review of 167
Untreated and 74 Treated Patients with Special Reference to Early Diagnosis," *Jour. Pediat.*, 57:807-823.

Lindsay, D. R., and Allen, E. M.
1961. "Medical Research: Past Support, Future Directions,"
Science, 134:2017-2024.

Long, E. R.
1928. *A History of Pathology.* Baltimore: Williams & Wilkins
Co.
1940a. "The Decline of Tuberculosis, with Special Reference
to Its Generalized Form," *Bull. History of Med.*, 8:819-
843.
1940b. "Weak Lungs on the Santa Fe Trail," *Bull. History of
Med.*, 8:1040-1054.
1953. "Tuberculosis in Modern Society," *Bull. History of Med.*,
27:301-319.
1958. *The Chemistry and Chemotherapy of Tuberculosis.* Baltimore: Williams & Wilkins Co.

Long, E. R., *et al.*
1957. "Report of Ad Hoc Advisory Committee on BCG to the
Surgeon General of the United States Public Health
Service," *Amer. Rev. Tuberc.*, 76:726-731.

Lorber, J.
1949. "A Study of Ten Children after Treatment with Streptomycin for Tuberculous Meningitis," *Arch. Dis. Childhood*, 24:289-293.

1954. "The Results of Treatment of 549 Cases of Tuberculous Meningitis," *Amer. Rev. Tuberc.*, 69:13-25.
1956. "Current Results in Treatment of Tuberculous Meningitis and Miliary Tuberculosis," *Brit. Med. Jour.*, 1:1009-1011.

Louis, P. C. A.
1844. *Researches on Phthisis*. Trans. W. H. Walshe. 2d ed. London: Sydenham Society.

Lowell, A. M.
1960. *Tuberculosis in New York City, 1959*. New York: New York Tuberculosis and Health Assoc.
1962. *Tuberculosis in New York City, 1961*. New York: New York Tuberculosis and Health Assoc.

McDermott, W.
1960. "Antimicrobial Therapy of Pulmonary Tuberculosis," *Bull. World Health Org.*, 23:427-461. Also in: 1962. *Amer. Rev. Resp. Dis.*, 86:323-335.

McDermott, W., *et al.*
1947. "Streptomycin in the Treatment of Tuberculosis in Humans: (I) Meningitis and Generalized Hematogenous Tuberculosis," *Ann. Intern. Med.*, 27:769-822.

McDougall, J. B.
1949. Tuberculosis: A Global Study in Social Pathology. Edinburgh: E. & S. Livingstone, Ltd.

MacGregor, I. M.
1962. "Where We Stand in Scotland," in *Tuberculosis—Prevention and Control*. London: The Chest and Heart Association.

MacLeod, J.
1956. "Streptomycin and Dihydrostreptomycin," *Practitioner*, 176:5-13.

Magarão, M. F., *et al.*
1961. "The Problem of the Resistance of *M. tuberculosis* to Standard Drugs in Rio de Janeiro, Brazil," *Adv. Tuberc. Res.*, 11:193-213.

Marks, J.
1925. *Genius and Disaster*. New York: Adelphi Publications. Also: 1928. London: John Hamilton, Ltd.

Millin, S. G.
1933. *Cecil Rhodes*. New York: Harper & Bros.

Møllgaard, H.
   1924. *Chemotherapy of Tuberculosis.* Copenhagen: Nyt Nor-
   disk Forlag.
Moorman, L. J.
   1940. *Tuberculosis and Genius.* Chicago: Univ. of Chicago
   Press.
Munro, D. G. M.
   1926. *The Psycho-pathology of Tuberculosis.* London: Oxford
   Univ. Press.
Myers, J. A.
   1927. *Fighters of Fate.* Baltimore: Williams & Wilkins Co.
Nannotti, A.
   1893. "Osservazione clinica e ricerche sperimentali intorna alla
   influenza delle infiammazioni da streptococco nelle af-
   fecioni tuberculari" (abstracted), *Centr. Bakteriol.,* 14:
   601-602.
Nicolson, D. W.
   1943. *Twenty Years of Medical Research.* New York: Na-
   tional Tuberculosis Assoc.
Ochs, C. W.
   1962. "The Epidemiology of Tuberculosis," *Jour. Amer. Med.
   Assoc.,* 179:247.
Perla, D. and Marmorston, J.
   1941. *Natural Resistance and Clinical Medicine.* Boston: Little,
   Brown & Co.
Pfuetze, K. H., *et al.*
   1955. "The First Clinical Trial of Streptomycin in Human
   Tuberculosis," *Amer. Rev. Tuberc.,* 71:752-754.
Piery, A. M., and Roshem, J.
   1931. *Histoire de la tuberculose.* Paris: G. Dion & Cie.
Porot, M.
   1950. *La psycholgie des tuberculeux.* Paris: Delachaux &
   Niestle.
Prasad, B. G.
   1961. "Pulmonary Tuberculosis in India," *Brit. Jour. Dis. Chest,*
   55:169-184.
Puig, C. X.
   1963. Viejos y nuevos problemas de la tuberculosis," *Inst. Antituberculoso "Francisco Moragas,"* 15:61-91.

Rappin, M.
    1912. "Action exercée par les diastases de quelques espèces sap-rophytes sur la tuberculose experimentale du cobaye," *Gaz. Méd. de Nantes,* pp. 958-960.
Rapport, S., and Wright, H.
    1952. *Great Adventures in Medicine.* New York: Dial Press.
Remarque, E. M.
    1961. *Heaven Has No Favorites.* New York: Harcourt, Brace.
Rich, A. R.
    1951. *The Pathogenesis of Tuberculosis.* 2d ed. Springfield, Ill.: Charles C. Thomas.
Riggins, H. M., and Hinshaw, H. C. (eds.)
    1949. *Streptomycin and Dihydrostreptomycin in Tuberculosis.* New York: National Tuberculosis Assoc.
Riley, R. L., *et al.*
    1962. "Infectiousness of Air from a Tuberculosis Ward," *Amer. Rev. Resp. Dis.,* 85:511-525. Also: 1962. *Adv. Tuberc. Res.,* 12:150-190.
Rolleston, J. D.
    1941. "The Folk-lore of Pulmonary Tuberculosis," *Tubercle,* 22:55-65.
Schatz, A., Bugie, E., and Waksman, S. A.
    1944. "Streptomycin, a Substance Exhibiting Antibiotic Activity against Gram-Positive and Gram-Negative Bacteria," *Proc. Soc. Exper. Biol. Med.,* 55:66-69.
Schatz, A., and Waksman, S. A.
    1944. "Effect of Streptomycin and Other Antibiotic Substances upon *Mycobacterium tuberculosis* and Related Organisms," *Proc. Soc. Exper. Biol. Med.,* 57:244-248.
Shattuck, L.
    1850. Report of the Sanitary Commission of Massachusetts. (Reprinted Harvard University Press, 1948.)
Shryock, R. H.
    1957. *National Tuberculosis Association, 1904-1954: A Study of the Voluntary Health Movement in the United States.* New York: National Tuberculosis Assoc.
Sjöberg, J. E., and Tivenius, L.
    1961. "Late Results of Thoracoplastics in Combination with Chemotherapy in the Treatment of Tuberculosis," *Acta. Tuberc. Scand.,* 40: 202-214.

Smith, D. T.
    1953.   "The Broad View: The Tuberculosis Problem in the
             United States," in National Tuberculosis Association,
             Transactions, 49th Annual Meeting, pp. 30-34.
Smith, G.
    1941.   *Plague on Us.* New York: The Commonwealth Fund.
Soper, F. L.
    1962.   "Problems To Be Solved If the Eradication of Tubercu-
             losis Is To Be Realized," *Amer. Jour. Pub. Health,* 52:
             734-745. Also: 1962. *Bull. Nat. Tuberc. Assoc.,* June, pp.
             9-12.
Stern, B. J.
    1941.   *Society and Medical Progress.* Princeton: Princeton Univ.
             Press.
Tellesson, W. G.
    1960.   "Changing Concepts in the Treatment of Tuberculosis
             (Part I)," *Med Jour. Australia,* 2:1005-1007.
Tillett, W. S.
    1948.   "The Antibiotic Age," editorial, *Amer. Jour. Med.,* 4:159-
             162.
Truant, J. P., Brett, W. A., and Thomas, W., Jr.
    1962.   "Fluorescence Microscopy of Tubercle Bacilli Stained with
             Auramine and Rhodamine," *Henry Ford Hosp. Med
             Bull.,* 10:287-296.
Trudeau, E. L.
    1916.   *An Autobiography.* New York: Doubleday, Page & Co.
             Also: 1928. New York: National Tuberculosis Assoc.
Vaudremer, A.
    1913.   "Action de l'extrait filtré d'*Aspergillus fumigatus* sur les
             bacilles tuberculeux," *Compt. Rend. Soc. Biol.,* 74:278-280.
Villemin, J. A.
    1865.   "Cause et nature de la tuberculose: Son inoculation de
             l'homme au lapin," *Compt. Rend. Acad. Sci.,* 61:1012-
             1015.
    1868.   *Études sur la tuberculose.* Paris: J. B. Ballière.
Waksman, S. A.
    1945.   "Planned Research in Biology, with Special Reference to
             Antibiotic Substances," Hearings on Science Legislation
             before the Subcommittee of the Committee on Military
             Affairs, U.S. Senate, 79th Congress, November 1-2.

1947. "Antibiotics and Tuberculosis," *Jour. Amer. Med. Assoc.,* 135:478-485.

1954. "Tenth Anniversary of the Discovery of Streptomycin, the First Chemotherapeutic Agent Found To Be Effective against Tuberculosis in Humans," *Amer. Rev. Tuberc.,* 70:1-8.

Waksman, S. A., Bugie, E., and Schatz, A.

1944. "Isolation of Antibiotic Substances from Soil Microörganisms, with Special Reference to Streptothricin and Streptomycin," *Proc. Staff Meet. Mayo Clinic,* 19:537-548.

Waksman, S. A., and Curtis, R.

1916. "The Actinomyces of the Soil," *Soil Sci.,* 1:99-134.

Waksman, S. A., and Henrici, A. T.

1943. "The Nomenclature and Classification of the Actinomycetes," *Jour. Bacteriol.,* 46:337-341.

Waksman, S. A., and Woodruff, H. B.

1942. "Streptothricin, a New Selective Bacteriostatic and Bactericidal Agent, Particularly Active against Gram-Negative Bacteria," *Proc. Soc. Exper. Biol. Med.,* 49:207-20.

Waring, J.J.

1953. "On Receipt of the Trudeau Medal," in National Tuberculosis Association, *Transactions, 49th Annual Meeting,* pp. 3-5.

Webb, G. B.

1936. *Tuberculosis.* New York: Paul B. Hoeber.

Weber, J. C. P., *et al.*

1963. "Long-Term Antibacterial Treatment for Pulmonary Tuberculosis in Patients Older than 5o Years," *Amer. Rev. Resp. Dis.,* 87:116-119.

Wilson, J. L.

1954. "The Future of Tuberculosis," in *Quarterly Progress Report of the Veterans Administration—Army-Navy Study on the Chemotherapy of Tuberculosis,* October. Washington, D.C.

Wolff, G.

1938. "Tuberculosis and Civilization: I. Basic Facts and Figures in the Epidemiology of Tuberculosis," *Human Biology,* 10:106-123.

1938. "Tuberculosis and Civilization: II. Interpretation of the

Etiological Factors in the Epidemiology of Tuberculosis,"
Human Biology, 10:251-284.

Woodruff, H. B., and Foster, J. W.
1944. *"In vitro* Inhibition of Mycobacteria by Streptothricin,"
Proc. Soc. Exper. Biol. Med., 57:88-89.

Yeager, R. L.
1962. "TB State of the Union—1962," *Nat. Tuberc. Assoc.
Bull.,* July-August, pp. 3-5.

Zorini, A. O.
1962. "Inaugurazione del 2° cours d'epidemiologie et de lutte
contre la tuberculose pour medecins Italiens et etrangers,"
Ann. dell'Istituto Carlo Forlanini, 22:3-10.

Zorzoli, G.
1940. "Influenza dei filtrati di alcuni miceti sul Bacillo tubercolare umano e bovino," *Ann. dell'Instituto Carlo Forlanini,* 4:208-220, 221-237.

INDEX

9 780520 328464